River Mechanics

Dai Wenhong Ana da Silva

·北京·

Abstract

The book *River Mechanics* was part of CIVL-455 (a graduate course of River Engineering at the Department of Civil Engineering, Queen's University, Canada) by Professors M. S. Yalin and Ana da Silva. This book includes four chapters: The first chapter introduces the basic knowledge of River Morphology, the second chapter introduces the Stabilization of River Course, the third chapter introduces the Modification of River Course, the last chapter introduces Special Problems in River Course, which introduces the discipline of river dynamics in detail. The purpose is to help readers understand the development law of the river and use its law to serve mankind, or guide it to develop in a direction conducive to mankind, or minimize the negative effects caused by changing the natural process of the river.

This book can be used as a textbook for college students majoring in water conservancy and environment, as well as a reference book for relevant researchers or engineering technicians.

图书在版编目（CIP）数据

河流动力学 = River Mechanics / 戴文鸿，（加）安娜·达席尔瓦著. -- 北京：中国水利水电出版社，2021.12
　　ISBN 978-7-5226-0116-8

Ⅰ.①河… Ⅱ.①戴… ②安… Ⅲ.①河流-流体动力学 Ⅳ.①TV143

中国版本图书馆CIP数据核字(2021)第209453号

书　　名	**River Mechanics（河流动力学）** River Mechanics（HELIU DONGLIXUE）
作　　者	Dai Wenhong ［加］Ana da Silva 著
出版发行	中国水利水电出版社 （北京市海淀区玉渊潭南路1号D座　100038） 网址：www.waterpub.com.cn E-mail：sales@waterpub.com.cn 电话：（010）68367658（营销中心）
经　　售	北京科水图书销售中心（零售） 电话：（010）88383994、63202643、68545874 全国各地新华书店和相关出版物销售网点
排　　版	中国水利水电出版社微机排版中心
印　　刷	天津嘉恒印务有限公司
规　　格	170mm×240mm　16开本　6.5印张　122千字
版　　次	2021年12月第1版　2021年12月第1次印刷
印　　数	0001—1000册
定　　价	**45.00元**

凡购买我社图书，如有缺页、倒页、脱页的，本社营销中心负责调换

版权所有·侵权必究

FORWORD

River is the product of the interaction between water flow and river bed under the influence of natural factors and human activities. River follows its own development law. On one hand, river flow acts on the river bed, causing the river bed to change and the river channel to evolve; On the other hand, the river bed also acts on the flow and affects the characteristics of the river. The two constitute a contradictory unity, interdependent, influencing and restricting each other, and are always in the process of change and development. River dynamics is a subject that studies the laws and applications of river flow, sedimentation and river bed evolution.

This book qualitatively and quantitatively describes the motion characteristics and development law of rivers. In addition to researchers and college students in the fields of hydraulic engineering, water resources and related geoscience branches, this book can also be used as a reference book for engineers and technicians.

This book includes four chapters: The first chapter introduces the basic knowledge of River Morphology, The second chapter introduces the Stabilisation of River Course, The third chapter introduces the Modification of River Course, The last chapter introduces Special Problems in River Course, which introduces the discipline of river dynamics in detail.

This book was edited by Professor Dai Wenhong of Hohai University and Professor Ana da Silva of Queen's University of Canada. It is also part of the graduate program of river engineering in the Department of civil engineering, Queen's University of Canada, by Professor M. S. Yalin and Professor Ana da Silva.

Due to the limited level of editors, this book will inevitably have fallacies and improper places. I sincerely hope that readers can criticize and correct it in order to improve in the future.

CONTENTS

FORWORD

INTRODUCTION 1

Chapter 1   RIVER MORPHOLOGY 3

   1.1   General 3

   1.2   Overall Erosion-Deposition Process (along $0<l<L$) 4

   1.3   Overall Development of River (along $0<l<L$) 6

   1.4   Flow Rate 10

Chapter 2   STABILISATION OF RIVER COURSE 21

   2.1   Prevention of Erosion 21

   2.2   Loose Revetment 29

   2.3   Prevention of Deposition 40

Chapter 3   MODIFICATION OF RIVER COURSE 48

   3.1   General 48

   3.2   Regime Channels 52

   3.3   Curvilinear River Regions 60

Chapter 4   SPECIAL PROBLEMS 71

   4.1   Flood Protection (Garde et al., 1977) 71

   4.2   Degradation and Aggradation (Yalin et al., 1983) 78

   4.3   Mathematical Formulation 81

REFERENCES 91

INTRODUCTION

"From the dawn of human history, rivers have been of high importance in the life and activities of man. The most ancient civilizations were built along the valleys of great rivers: Mesopotamian civilization along the Tigris and Euphrates, Indian along the Ganges and Indus, Egyptian along the Nile. The Yangtze River was of great significance to the development of China; the Rhine and Danube to the development of Europe (Morris et al., 1963)." Rivers supply water (for biological, sanitary and agricultural use), they provide food (fish), they open the possibility for the production of electrical power and transport ("waterways" in addition to "highways"). On the other hand, rivers may appear as the sources of various damages. Rivers can cause catastrophic floodings in populated and agricultural areas, it may induce the landslides; its plane shape may not be stable and its displacement in plane ("river-wandering") can create a number of difficulties (Makkaveev et al., 1955). It is thus clear that man cannot remain indifferent to the river: he is compelled to take actions to increase the benefits and to reduce the damages. And this is how the River Engineering (one of the oldest engineering activities) came into being. Unfortunately, this interesting discipline cannot be studied in detail here and the present text is but an outline of some of it's basic principles and methods.

The main objectives of River Engineering can be summarized as follows (Jansen et al., 1979; Garde et al., 1977; Duhm et al., 1951; Grishin et al., 1955):

INTRODUCTION

(1) Stabilisation of river course that is the prevention of its <u>deformation</u> and/or <u>displacement</u> ("river wandering"). (These tendencies are due to erosion and/or deposition processes).

(2) Modification to river course (as to suit best to agriculture, land reclamation, navigation, and other socio - economic requirements).

(3) Handling of special problems caused by river Flood protection works, aggradation and degradation of the river bed (e.g. due to an erected dam), changes in the river regime due to the construction of diversion channels (fed by that river), etc.

Before entering the study of these topics, however, one must acquire first some knowledge on River Morphology which forms the "starting point" of the present course.

I am grateful to my friend Professor K. C. Wilson for his constructive comments and suggestions.

Chapter 1 RIVER MORPHOLOGY

1.1 General

River Morphology, concerns the "general form and behavior of a river as a whole" (Jansen et al., 1979). There are "identical rivers"; and yet most of the rivers exhibit a number of common features and trends. Some of them (those which are relevant to the River Engineering) are outlined in this section. The pertinent terminology related to a river course is given in Fig. 1.1.

As can be inferred from Fig. 1.1, a river usually is but a "main stream" of a "system of streams" (Jansen et al., 1979). As a rule: River source is either a spring (Amazon, Danube, Volga, etc.) or a lake (St. Lawrence, White Nile, Mississippi, etc.) or a glacier (Athabasca, North Saskatchewan, Rhone, etc.) River-end is either a "River mouth" (St. Lawrence, Colorado, Thames, etc.) or a "Delta" (Mackenzie, White Nile, Amazon, Mississippi, etc.)

In the following, the location of a river section will be specified by its distance l to the source. Thus source implies $l=0$, and if river-end is denoted by $l=L$, then for l we have the range: $0 < l < L$. Most of the quantitative properties of a river exhibit a substantial variation with the distance l, and some of the properties vary also significantly with the time t. Thus, in general, a quantitative property is A, a river can be considered as

Chapter 1 RIVER MORPHOLOGY

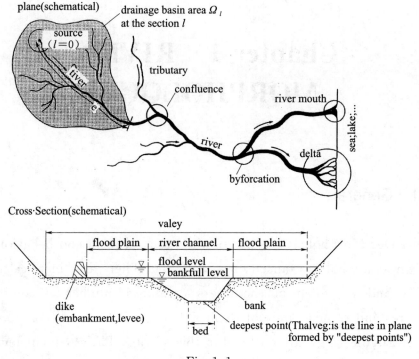

Fig. 1. 1

$$A = f_A(l, t) \tag{1.1}$$

At a given river section (l=const):

$$A = f_A(\text{const}, t) = \phi_A(t) \tag{1.2}$$

At a given instant (t=const):

$$A = f_A(l, \text{const}) = \psi_A(l) \tag{1.3}$$

(e. g. the drainage basin area Ω_l varies significantly with l but not with t, whereas the flow rate Q usually varies strongly with both l and t.)

1.2 Overall Erosion – Deposition Process (along $0 < l < L$)

At a given t the flow depth h and the slope S vary along l as

1.2 Overall Erosion – Deposition Process (along $0<l<L$)

shown schematically in Fig. 1.2.

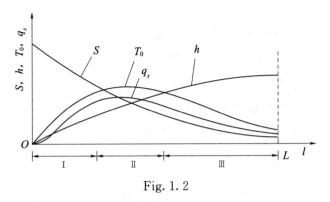

Fig. 1.2

But if so, then bed shear stress $\tau_0 \approx \gamma h S$ must vary as indicated by the curve τ_0 while the transport rate per unit flow width q_s (which is an increasing function of τ_0) as implied by the curve q_s. The character of the curve q_s suggests that the following three typical zones must be present along $0<l<L$

$$\left. \begin{array}{l} \text{zone I where } (\partial q_s/\partial l)>0 \\ \text{zone II where } (\partial q_s/\partial l)=0 \\ \text{zone III where } (\partial q_s/\partial l)<0 \end{array} \right\} \quad (1.4)$$

Using equation (1.4) in the Exner – Polya equation

$$\frac{\partial q_s}{\partial l}+\gamma_s \frac{\partial z}{\partial t}=0 \qquad (1.5)$$

(where $q_s=f_{q_s}(l, t)$ and $z=f_z(l, t)$)

one arrives at the conclusion that

In the zone I : $(\partial z/\partial t)<0$, i.e. bed elevation z decreases with time (erosion in the "upper reach" of a river);

In the zone II : $(\partial z/\partial t)=0$, i.e. bed elevation z hardly varies with t (equilibrium in the "middle reach" of a river);

In the zone III : $(\partial z/\partial t)=0$, i.e. bed elevation z increases with time (deposition in the "lower reach" of a river).

Chapter 1　RIVER MORPHOLOGY

These trends are illustrated schematically in Fig. 1. 3.

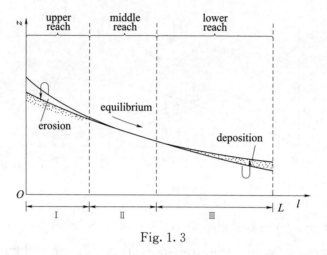

Fig. 1. 3

1.3　Overall Development of River (along $0 < l < L$)

With the increment of l, the size D of the transported solids decreases: rocks turn into gravel and then into the sand, and sand becomes finer and finer (Fig. 1. 4).

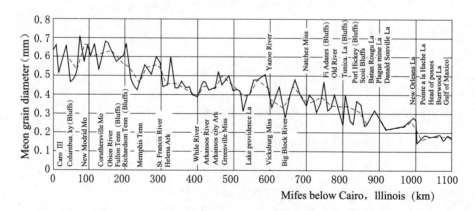

Fig. 1. 4

On the other hand, the increment of l means the "movement" from a mountainous region towards the sea and thus it means the

· 6 ·

1.3 Overall Development of River (along $0 < l < L$)

widening and flattening of the river valley (the "V" - the shape of its cross-section progressively turns into a wide "U" - shape). The sand deposited in the zone Ⅲ (alluvial deposition) usually covers the whole width of the valley (due to the spills during floods and also due to the lateral "wanderings" of river channel throughout the periods of its geological evolution). Thus, in the lower reaches, river channel usually is situated completely in alluvium (it is an "alluvial channel"). The developments outlined above are illustrated schematically in Fig. 1.5.

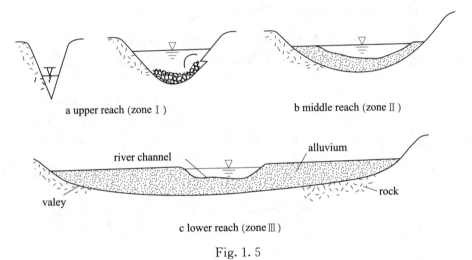

a upper reach (zone Ⅰ)   b middle reach (zone Ⅱ)

c lower reach (zone Ⅲ)

Fig. 1.5

From an engineering standpoint rock forms a rigid flow boundary, whereas alluvium is a deformable boundary. (From geological standpoint, where duration is measured by tens or hundreds of millenniums, even the hardest rock cannot be regarded as "rigid": formation of canyons). Accordingly, in the rocky upper reaches the plane shape of the river is determined by that of the valley, whereas in the alluvial lower reaches it is determined by the river flow itself (which carves its channel in alluvium "as it wants" and thereby generates "its own" typical plane patterns). When S is sufficiently large then the alluvial river channel usually is reasonably

Chapter 1   RIVER MORPHOLOGY

straight in plane. With the decrement of S this (single) channel tends to split into a multitude of irregular curvilinear channels (braiding). A further decrement of S leads again to the formation of a single channel. But this time the channel is no longer straight: it exhibits in plane a serpent - like form (meandering). Since decrement of S is usually accompanied with the increment of l (Fig. 1. 2), as a rule, for an alluvial channel the following sequence of development along l:

The alluvial channel changes from straight state to braiding state and finally develops into meandering channel.

This sequence is illustrated schematically in Fig. 1. 6. (Observe, that Fig. 1. 5c can be regarded thus as a section of Fig. 1. 6).

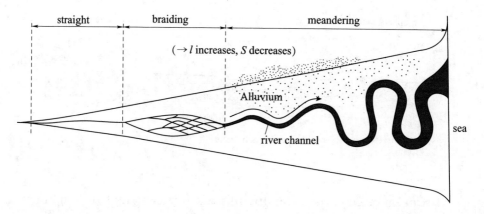

Fig. 1. 6

A real - river example is shown in Fig. 1. 7 (Jansen et al., 1979). Observe from the longitudinal section that the slope of the Tigris River suddenly changes at the town Balad ($l = 367$km). For $l < 367$km the slope S is "large" and the river is braiding. For $l \geqslant 367$km the slope is "small" and the river is meandering.

Note: For better visibility a river cross - section usually is pictured with an exaggerated depth to width ratio: e. g. as in Fig. 1. 8a

1.3 Overall Development of River (along $0<l<L$)

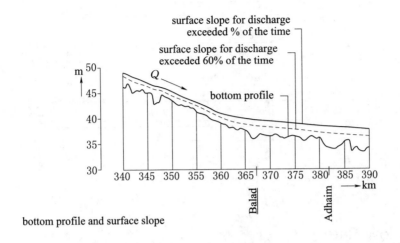

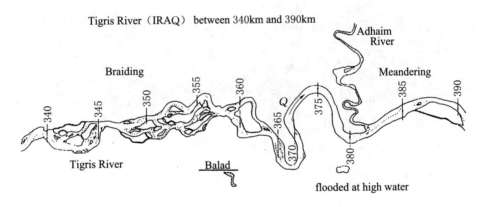

Fig. 1.7

where h/B is 1/5. The "optical illusion" conveyed by the sketches like this led to a number of misconceptions in River Mechanics. In fact, the order of h/B for the majority of practical cases is more like 1/50, or even less (especially for the rivers in alluvium), and therefore the actual shape of a river cross-section is more likely to be as in Fig. 1.8b (Jansen et al., 1979). Considering this, in the following we will frequently take $B/B_0 \approx 1$ and we will treat the ratio h/B as "small".

· 9 ·

Chapter 1 RIVER MORPHOLOGY

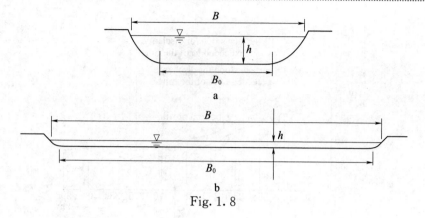

Fig. 1. 8

1.4 Flow Rate

(1) At a given instant t, the general form $Q=f_Q(l, t)$ reduces to

$$Q=\phi_Q(l) \tag{1.6}$$

Fig. 1. 9 shows the equation (1. 6) for the River Danube, here Q_{av} is the mean annual-average flow rate (which will be defined later on).

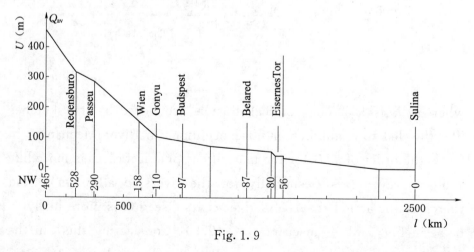

Fig. 1. 9

(2) At a given river section l, the general form $Q=f_Q(l, t)$ reduces to

1.4 Flow Rate

$$Q = \phi_Q(t) \qquad (1.7)$$

The equation (1.7) is more relevant to River Engineering than equation (1.6) and is elaborated below. Fig. 1.9 is called the "hydrograph" (of the river at that l). The character of a hydrograph varies from one river to another (depending on climate conditions and geomorphological conditions), and for a given river it is not the same every year. Note e. g. from Fig. 1.10a and Fig. 1.10b that the peaks of the Rhine river do not occur in as predictable times of the year as in the case of the Athabasca River (June, July) nor the magnitude of the ratio Q_{max}/Q_{min} of the Rhine is as large as that of the Athabasca (different character). Note also that in both of these rivers the hydrographs are far from being identical every year.

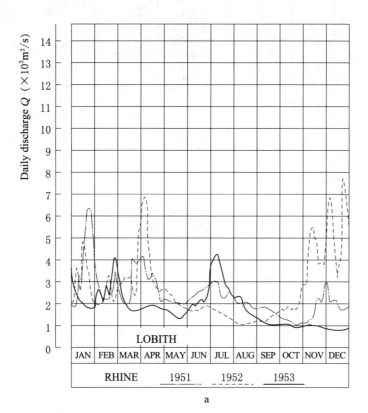

Fig. 1.10 (A)

Chapter 1 RIVER MORPHOLOGY

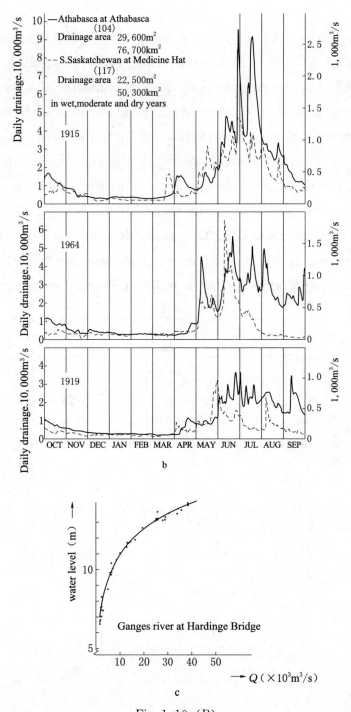

Fig. 1.10 (B)

1.4 Flow Rate

The time variation of Q is accompanied by the time variation of water level (WL). The values of Q and WL corresponding to the same section l are interrelated by the "Rating Curve" (Fig. 1.10c)

It is supposed that at a river – section l the values of Q were measured every day for a large number (N) of days. In the following we will associate <u>one</u> day with <u>one</u> value of Q: namely with the <u>largest</u> Q that occurred that day (Leopold et al., 1964). Furthermore, we will use the notation N_x to designate the number of days when Q is larger than a prescribed possible value x.

The sequence of Q – values measured in a year j appears as in Fig. 1.11a. This sequence, which is somewhat above the hydrograph $Q = \phi_Q(t)$, is obviously correlated (for the Q_i – value corresponding to a day i is likely to be near to Q_{i-1} corresponding to the day $i - 1$). However, an engineer is not concerned with the exact date of occurrence of a particular Q; nor is he interested in the value of flow rate a day before that date. The question that is of interest to him is "what is the frequency N_x/N of the occurrence of the flow rates which exceed a particular value x (and which may thus endanger his project)?" But the answer to this question would still be the same even if the sequence of "vertical rectangles" in Fig. 1.11a underwent a "random shuffling"; i.e. if the correlated sequence in Fig. 1.11a were replaced by the uncorrelated (and thus mathematically more convenient) random sequence in Fig. 1.11b.

The random sequence in Fig. 1.11b can be regarded as the realization (sample: year i) of a discrete stationary stochastic process $Q_i = \phi_*(t_i)$, say. The total number of all the available "vertical rectangles" (such as those forming Fig. 1.11b) is N. Organizing them in increasing order and rendering the time axis dimensionless by dividing each "day" by "N days" (and thus by using unity instead of "N days" and l/N instead of "one day") one arrives at a curve such

Chapter 1 RIVER MORPHOLOGY

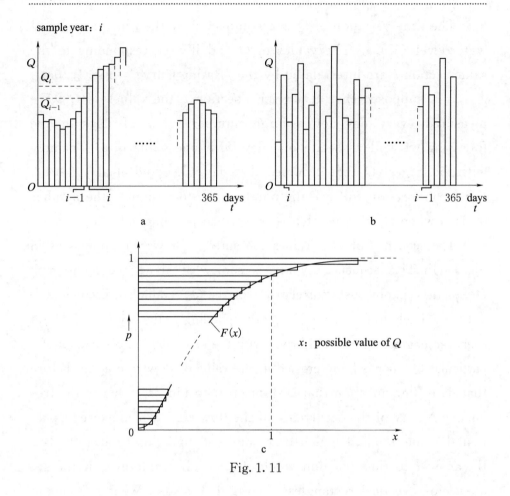

Fig. 1.11

as $F(x)$ shown in Fig. 1.11c.

$$F(x) = \frac{N_x}{N} = [\% \text{of days with } Q<x] = \text{Prob}(Q<x) = p \quad (1.8)$$

and thus $F(x)$ is the probability distribution function (of the random process $Q_i = Q_*(t_i)$). Some realistic examples of $F(x)$ are shown in Fig. 1.12.

Equation (1.8) from unity one obtains the probability $P = 1-p$ of the complimentary event $Q>x$:

$$1-F(x) = \frac{N-N_x}{N} = [\% \text{ of days wigh } Q>x] = \text{Prob}(Q>x) = p$$

$$(1.9)$$

1.4 Flow Rate

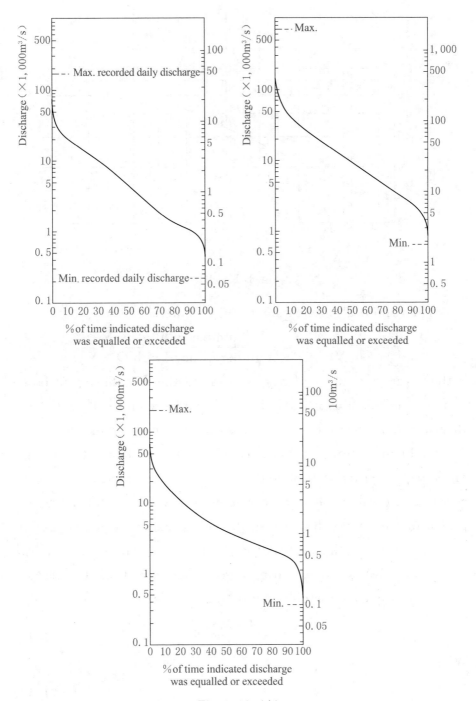

Fig. 1.12 (A)

Chapter 1 RIVER MORPHOLOGY

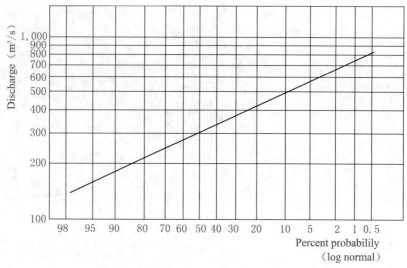

Fig. 1.12 (B)

The reciprocal of equation (1.9), namely

$$r = \frac{1}{\bar{p}} = \frac{N}{N-N_x} = \frac{[\text{number of all days}]}{[\text{number of days with } Q<X]} \quad (1.10)$$

is the "average number of days separating the days having Q larger than a specified value x" or it is the "recurrence interval" of $Q>x$. Recurrence interval must not necessarily be measured in "days" (as one might infer from equation (1.10)), and in fact, it is usually measured in "years". It should also be borne in mind that "the recurrence interval is not a forecast: a flood which would not be expected to occur more than once in 100 years, might occur next year". But what then is the probability P_M that a Q larger than x will occur (exactly) M times during a given period of time T? The answer to this question is given by the Poisson formula

$$P_M = \frac{\bar{P}TM}{M!} e^{-\bar{P}T} \quad (1.11)$$

Where

$$\bar{p} = 1 - p = \frac{1}{r} \quad (1.12)$$

(T and r must be expressed in terms of the same time unit).

1.4 Flow Rate

The following special values of Q are pertinent to the present course. (All of these Q_S correspond to a specified River Section 1; if N is the number of days (for which Q was measured) then the number of years is $n = N/365$).

(a) Mean annual average flow rate $\overline{Q}_{av}$: is given by

$$\overline{Q}_{av} = \frac{1}{n}\sum_{j=1}^{n}(Q_{av})_j \qquad (1.13)$$

where $(Q_{av})_j$ is the average flow rate of the year j.

(b) Mean annual maximum flow rate (or "mean annual flood") $\overline{Q}_{max}$:

$$\overline{Q}_{max} = \frac{1}{n}\sum_{j=1}^{n}(Q_{max})_j \qquad (1.14)$$

where $(Q_{max})_j$ is the maximum flow rate of the year j.

(c) Bank full flow rate Q_{bf}:

The meaning is clearly express from Fig. 1.1. According to the prevailing view, it is Q_{bf} that determines the geometry of the river channel in alluvium.

(d) "M - year flood" Q_M: is the flow rate that corresponds to a specified recurrence period r. e. g. "100 - year flood" Q_{100} is the flowrate that occurs, on the average, once in 100 years (which corresponds to $M = 100$ years). River Engineering structures are designed so as to withstand the forces caused by a certain adopted flow rate Q_M (usually Q_{100}) (Garde et al., 1977; Morris et al., 1963).

The flow rates defined above, as a rule, compare with each other as

$$\overline{Q}_{av} < Q_{bf} < \overline{Q}_{max} < Q_M \qquad (1.15)$$

According to (Leopold et al., 1964) the recurrence intervals of some Q vary (from one river to another) within remarkably narrow ranges. Thus

$$\text{for } \overline{Q}_{max}: M = 2 \text{ to } 2.5 \text{ yr } (M_{av} = 2.33\text{yr}) \qquad (1.16)$$

$$\text{for } Q_{bf}: \text{we have } M = 1 \text{ to } 2 \text{ yr} \qquad (1.17)$$

Chapter 1 RIVER MORPHOLOGY

These ranges of M were obtained from measurements of (mostly) U. S. rivers. For British rivers, Q_{bf} appears to correspond to $M=0.5$ yr (Henderson et al., 1966; Nixon et al., 1959). However, this (substantial) difference is very likely to be due to the different methods of measurement adopted in Nixon et al. (1959) and Leopold et al. (1964).

Some realistic examples for the variation of Q (or a dimension-

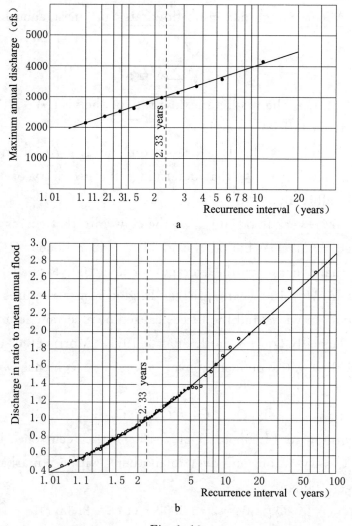

Fig. 1.13

1.4 Flow Rate

less version thereof) with M are shown n in Fig. 1.13. ("Large" values of Q are not restricted. "Small" values of Q are restricted by the fact that they cannot be negative. Hence a certain "skewness" is to be expected in the graphs of Q versus M or p. Owing to this reason log-normal plots are often utilized (Fig. 1.12 and Fig. 1.13). See more on log-normal plots e.g. in Viessman et al. (1977).

" The total annual discharge in the rivers of the world is about 8,200 cubic miles, representing about one-third of the annual precipitation (Morris et al., 1963)." The mean annual average flow rates $\overline{Q}_{av}$ (at the River end $l=L$) of the ten largest rivers are shown in Table 1.1.

Table 1.1 (Morris et al., 1963)

Rank	River	Average Discharge (cfs)
1	Amazon (Brazil)	4,000,000
2	LaPlata-Parana (Argentina)	1,600,000
3	Congo (Africa)	1,400,000
4	Yangtze (China)	1,000,000
5	Ganges-Brahmaputra (India)	800,000
6	Mississippi (United States)	620,000
7	Yenisei (Siberia)	610,000
8	Orinoco (Venezuela)	600,000
9	Mekong (Thailand)	560,000
10	Lena (Siberia)	540,000

Some characteristics of prominent Canadian Rivers are given in Fig. 1.14. (Here too the values of $\overline{Q}_{av}$ corresponding to $l=L$). See Bullock et al. (1965) to Edmonton et al. (1972).

Chapter 1 RIVER MORPHOLOGY

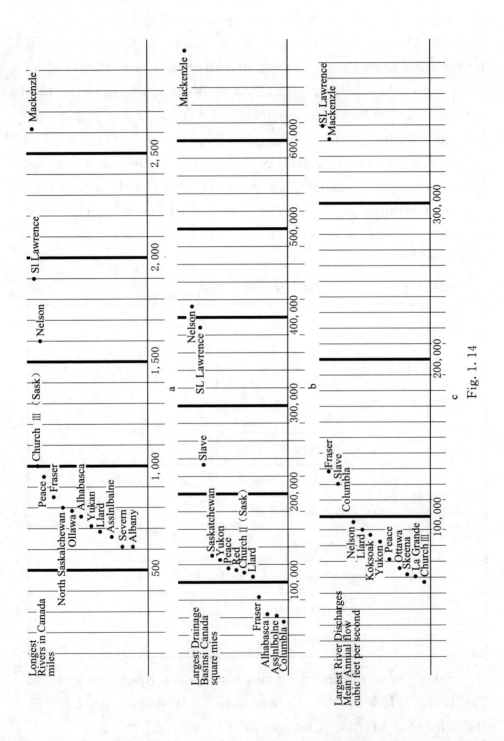

Fig. 1.14

Chapter 2 STABILISATION OF RIVER COURSE

2.1 Prevention of Erosion

In principle, erosion can be prevented

(1) by reducing the flow action on the endangered boundary (i. e. by reducing τ_0 and μ_0) without altering the material forming the (loose) flow boundary.

(2) by increasing the resistance of the endangered boundary (by covering it with a material (revetment) having sufficiently large $(\tau_0)_{cr}$); without altering the action of flow.

The combination of (1) and (2) can also be utilized.

2.1.1 Upper Reach (Zone I in Fig. 1.3)

Here the river (or any of its tributaries) manifests itself as a "mountain stream". (Typically: $\approx 1 \text{ km}^2 < \Omega_l < \approx 50 \text{km}^2$ and S up to ≈ 0.1; $\approx 1\text{cm} < D < \approx 20\text{cm}$ (some individual stones up to $D \approx 1\text{m}$); $\approx 2\text{m/s} < v < \approx 10\text{m/s}$ (in some cases, like the Fraser River, up to $v \approx 15\text{m/s}$); side slopes of the valley up to $\approx 1/1$ or even steeper. up to $2 \times 10^5 \text{N/m}^2$ (e. g. (Zamarin et al., 1952). As can be inferred from the content of 1.2 and 1.3 (Fig. 1.3 and Fig. 1.5), it is the river bed, rather than the rocky side walls (banks) that might be subjected to excessive erosion. Usually, the erosion of the bed of a mountain stream is prevented with the aid of the method (1)

Chapter 2　STABILISATION OF RIVER COURSE

which can be applied in the following two ways.

(1) Series of structures: Build a series of "sills" (Fig. 2.1a) or "steps" (Fig. 2.1b). The stream will "fill" or "excavate" the regions upstream of these structures. Thereby it will reduce S and thus $\tau_0 \sim S$ so that τ_0 becomes as low as $(\tau_0)_{cr}$ and the transport (erosion) will be terminated (Jansen et al., 1979; Garde et al., 1977; Duhm et al., 1951; Grishin et al., 1955) etc.

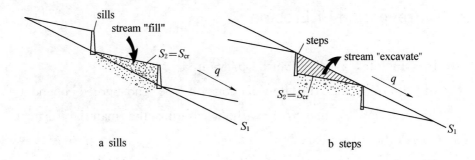

Fig. 2.1　Series of structures

In the case of "sills" river level is elevated, in the case of "steps", it is lowered. The elevation of the river leads to the increment of its width. Consequently, it leads to the reduction of the depth h and thereby to the further reduction of $\tau_0 \sim h$. Hence "sills", which are slightly more economic to build (less excavation), are also slightly more advantageous from the hydraulic standpoint. However, these (slight) advantages may sometimes lose their meaning (e. g. if the elevation of river water level and thus of the groundwater level of the neighboring terrain is for some reason undesirable). The (dotted) area upstream of the "sills" is filled by a stream; the (shaded) area upstream of the "steps" is excavated by a stream. In either case, the final slope of the bed will correspond to $(\tau_0)_{cr}$.

Denoting the initial and final characteristics by the subscripts 1 and 2, and treating the flow as two-dimensional one can write (ap-

2.1 Prevention of Erosion

proximately)

$$q = \frac{1}{n_1} h_1^{\frac{2}{3}} S_1^{\frac{1}{2}} = \frac{1}{n_2} h_2^{\frac{2}{3}} S_2^{\frac{1}{2}} \tag{2.1}$$

which assuming $n_1 = n_2$ gives

$$\frac{S_1}{S_2} = \left(\frac{h_2}{h_1}\right)^{\frac{4}{3}} \tag{2.2}$$

Furthermore, we have

$$Y_{cr} = \frac{1}{1.65} \frac{S_2 h_2}{D} \approx 0.05 \text{ and } H = (S_2 - S_1) L \tag{2.3}$$

equation (2.2) and equation (2.3) imply 3 independent equations involving 7 quantities: S_1, h_1, D, S_2, h_2, H and L. Knowing S_1, h_1, D and adopting a value e.g. for H one can determine the remaining 3 variates S_2, h_2, and L by solving 3 equation (2.2) and equation (2.3). If the result is unreasonable, adopt another value for H and repeat the procedure.

(2) Widening of river channel: Excavate the river channel sideways, thereby increase its width from B_1 to B_2 and thus reduce its depth from h_1 to h_2 so that τ_0 becomes as low as $(\tau_0)_{cr}$ and the transport (erosion) terminates (Fig. 2.2) (Jansen et al., 1979; Garde et al., 1977; Duhm et al., 1951; Grishin et al., 1955).

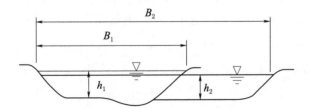

Fig. 2.2 Widening of river channel

Since

$$Q = \frac{1}{n_1} B_1 h_1^{\frac{5}{3}} S_1^{\frac{1}{2}} = \frac{1}{n_2} B_2 h_2^{\frac{5}{3}} S_2^{\frac{1}{2}} \tag{2.4}$$

one obtains (assuming $S_1 \approx S_2$ and $n_1 \approx n_2$)

Chapter 2　STABILISATION OF RIVER COURSE

$$\frac{h_1}{h_2} = \left(\frac{B_2}{B_1}\right)^{\frac{3}{5}} \tag{2.5}$$

Furthermore:

$$Y_{cr} = \frac{1}{1.65} \frac{S_1 h_2}{D} \approx 0.05 \tag{2.6}$$

Eliminating h_2 from equation (2.5) and equation (2.6) one determines for the required width B_2

$$B_2 = B_1 = \left[\frac{h_1 S_1}{1.65 \times 0.05 \times D}\right]^{\frac{5}{3}} \tag{2.7}$$

2.1.2　Lower Reach (Zone III in Fig. 1.3)

Here we have a "low-land river" flowing in alluvium. [Typically: $S < \approx 0.001$; $V \approx 2m/s$; $\approx 0.1mm < D < \approx 0.5mm$ (Leopold et al., 1964; Grishin et al., 1955; Makkaveev et al., 1955). As can be inferred from the content of 1.2 and 1.3 (Fig. 2.3 and Fig. 2.4), the river is in the zone of deposition, and therefore the erosion of its bed is not likely to be the case. On the other hand, it is exactly in this zone where the river creates "its own" plane geometry (braiding, meandering) which usually varies in time: river channel is displacing ("wandering") and deforming (widening). But since any displacement or widening of the river channel is inevitably associated with the continual erosion of (at least one of) its banks, one concludes that in lower reaches it is mainly the erosion of banks (rather than of the bed) that may form the concern. The prevention of bank erosion can be carried out either by means of the method (1) or by means of the method (2). We begin with (1).

(1) Deflection of Flow from Eroding Bank.

A river in the alluvium is usually serpent-like in plane

2.1 Prevention of Erosion

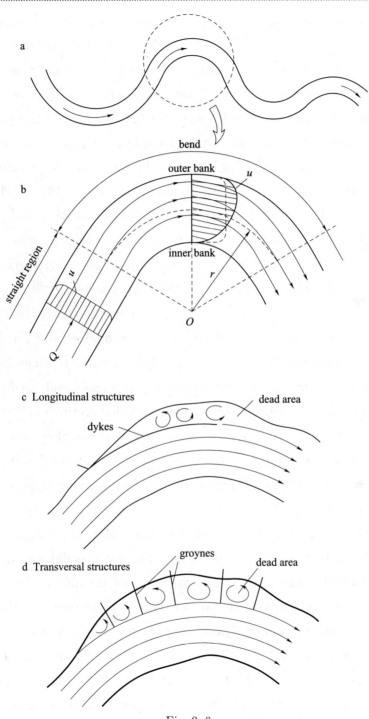

Fig. 2.3

Chapter 2 STABILISATION OF RIVER COURSE

(Fig. 2.3a). Consider one of its straight regions ("bend" in Fig. 2.3b). The centrifugal force "pushes" the bulk of the moving fluid mass off the center of rotation 0. Consequently, the flow velocities u at the outer (concave) bank are larger, while those at the inner (convex) bank are smaller than their counterparts of the symmetrical u – diagram corresponding to the straight river region (more information on open channel flow in bends can be found in (Rosovskii et al., 1957) and (Falcon et al., 1959; Zimmermann et al., 1978). Hence it is the outer bank (which is acted upon by "large" velocities u) that is often subjected to a continual erosion; the inner bank) where u are "small", is subjected to deposition (Jansen et al., 1979; Garde et al., 1977; Henderson et al., 1966; Rosovskii et al., 1957; Grishin et al., 1955, etc). In order to reduce the velocities u at the outer bank, and thereby to eliminate its continual erosion, it is sufficient to deflect the flow away from it. Such a deflection can be realized either by means of <u>longitudinal structures</u> referred to as <u>dykes</u> (Fig. 2.3c) or by means of <u>transversal structures</u> called <u>spurs</u> or <u>groynes</u> (Fig. 2.3d). After the erection of any of these structures the fluid area at the outer bank turns into a "dead area" containing slowly rotating vortices (Duhm et al., 1951; Grishin et al., 1955; etc). These (slow) vortices are not capable of producing erosion, and in fact, it is the deposition that usually takes place in the "dead area". If the river is (reasonably) straight but it is deforming (widening) then this can also be prevented by deflecting the flow from the bank (or banks) with the aid of dykes or groynes as outlined above. More information on dykes and groynes will be given in Section 2.3 (Jansen et al., 1979; Garde et al., 1977; Stephenson et al., 1977; Duhm et al., 1951; Grishin et al., 1955); for the present it is sufficient to infer that, in essence, they are merely some "guide vanes" or "deflectors" which can be used to alter the flow pattern.

2.1 Prevention of Erosion

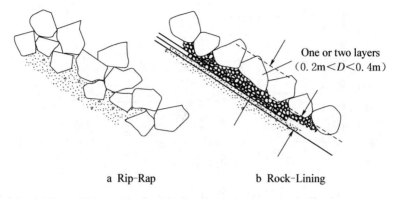

a Rip-Rap   b Rock-Lining

Fig. 2.4 Loose Revetment

(2) Bank Protection.

In this case, the bank subjected to Grishin erosion is covered by a protective material (revetment) that can withstand the shear action of the existing flow (Method (2)). A variety of materials can be used for this purpose and they are selected depending on the magnitude of shear action. To ensure flexibility, the protective coverings are frequently formed from various kinds of "mattresses" (e.g. woven willow mats, willow facine mats, reinforced asphalt mats, concrete, and reinforced concrete mats, etc.). Similarly, wire – mesh mattresses and boxes filled with stones "Reno mattresses" and "Gabions"), which can withstand the shear action up to $\tau_0 = 10$ to 20 kN/m², can be utilized (Fig. 2.5). The banks of rivers passing through the large cities are often protected by concrete, reinforced concrete, or cemented stone layers (Duhm et al., 1951; Press et al., 1956; Grishin et al., 1955; etc). These (costly) revetments are used, however, only very seldom in rural areas.

An effective and yet economic method for bank protection, which is extensively used nowadays in all areas, is the loose revetment which consists of stones or rocks ($0.2\text{m} < D < \approx 0.4\text{m}$) bedded on the river bank surface (Fig. 2.5). Stones forming the

Chapter 2 STABILISATION OF RIVER COURSE

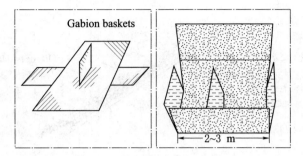

Fig. 2.5

loose revetment are either simply dumped on the bank (rip‐rap) or they are placed on it (rock‐lining) (Jansen et al., 1979; Garde et al., 1977; Stephenson et al., 1977; Grishin et al., 1955; etc). (To prevent the "washing out" of the (fine) alluvial bank material through the spaces between the stone forming the rock‐lining "and yet to allow the passage of ground water" (the bank surface usually is covered first by a "filter", that is by a layer of graded gravel or quarry chip (Jansen et al., 1979; Garde et al., 1977; Stephenson et al., 1977; etc).) The size of stones forming the loose revetment must be determined based on mechanical considerations presented in the next section.

2.1.3 Middle Reach (Zone II in Fig. 1.3)

Although in this zone the river is expected to be in equilibrium some deviations from it are only natural under practical circumstances. Thus in some locations, an excessive erosion can be present and one may wish to prevent it. Since the zone II is an "intermediate" (or "transitional") zone between the zones I and III the erosion of both bed and banks (which are typical for the zones I and III respectively) can be expected (Duhm et al., 1951; Grishin et al., 1955). Here, the prevention of erosion of banks is carried out by using the same methods as in the lower reaches (i.e. as outlined in

2.1.2). Yet the bed erosion usually is prevented, as shown in Fig. 2.6, that is, with the aid of a series of "sills" built on the endangered bed surface (rather than as outlined in 2.1.1). The sills slow down the flow at the bed (and thereby reduce τ_0 acting on the bed) by deflecting the stream lines away from the bed surface (Method (1)).

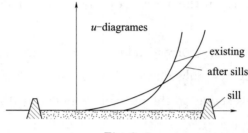

Fig. 2.6

2.2 Loose Revetment

2.2.1 General

(1) So far we were dealing with movable beds having "small" angles of inclination θ (indeed, we were dealing with slopes $S = \tan\theta$ such as 0.001, 0.01, etc). Now, we will consider also movable <u>banks</u> having finite angles θ (e.g. $\tan\theta = 1/2$, $1/3$, $1/1.5$, etc).

(2) In this section, we will deal only with the shear stresses <u>acting on flow boundaries</u> and we will denote them simply by τ (not by τ_0). The flow velocity <u>at the flow boundaries</u> will be denoted by U.

(3) Characteristics corresponding to loose boundaries having "small" ($\theta \rightarrow 0$) inclinations (such as river beds) will be marked in the following by the subscript "0" (e.g. τ_0, U_0, $(F_x)_0$, D_0,

etc); those corresponding to the loose boundaries having finite inclinations e will be marked by the subscript "θ" (τ_θ, U_θ, $(F_x)_\theta$, D_θ, etc) Characteristics which are equally valid for both "small" and finite values of θ will not be marked by "0" or "θ" (g, v, ρ, γ_s, etc). (Hence in the present section τ_0 means the "shear stress acting on the boundary having" small "inclination" and not the "shear stress acting on the surface of any flow boundary" (as it usually implies).)

(4) Characteristics corresponding to the critical stage (c - stage) will be marked by the subscript "c" (not by "c_r"): e.g. $\tau_{\theta c}$ is the critical shear stress acting on the movable boundary with finite inclination (acting on a loose bank).

(5) If θ is "small" then the detachment of grains is usually due to the lift force $\vec{F}_y$ and it is only in some cases that it is due to the drag force $\vec{F}_x$ (isolated grains situated higher than their "neighbors"). If θ is finite then the detachment is invariably due to the drag force $\vec{F}_x$ (White et al., 1940; Stephenson et al., 1977; Anderson et al., 1970; etc). The stones or rocks forming the loose revetment withstand the action of $\vec{F}_x$ by means of the friction force, friction coefficient f being equal to the tangent of the angle φ of repose.

In order to have a "unified approach" it will be assumed in this section that the detachment is always due to the drag force $\vec{F}_x$ irrespective of whether θ is finite or "small". We begin with the consideration of "small" θ.

2.2.2 Critical Stage of Loose Revetment for "Small" θ

As has been pointed out in the preceding section we have two types of loose revetment:

2.2 Loose Revetment

(1) Rip-Rap:

formed by dumped stones Surface roughness is larger than

$$k_s = (2 \text{ to } 3) D_0 \qquad (2.8)$$

The revetment fails if some exposed stones fail to withstand the action of $(\vec{F_x})_0$ which is determined by U_0 (Stephensonet al., 1977):

$$(F_x)_0 = C_1 \rho D_0^2 U_0^2 \qquad (2.9)$$

(2) Rock-Lining:

formed by hand-packed stones. Surface roughness is smaller than in (1):

$$k_s = (1 \text{ to } 2) D_0 \qquad (2.10)$$

The revetment fails if an area $(A_0 = C_2 D_0^2)$ of the rock layer fails to withstand the action of $(\vec{F_x})_0$ which is determined by τ_0 (Stephensonet al., 1977):

$$(F_x)_0 = C_2 D_0^2 \tau_0 \qquad (2.11)$$

The exposed stone will not be dislodged by the flow as long as

$$\frac{(F_x)_0}{G_0} < f = \tan\phi \qquad (2.12)$$

where G_0 is the submerged weight of the stone:

$$G_0 = \gamma_s \alpha_1 D_0^3 \qquad (2.13)$$

The rock layer will not be dislodged by the flow as long as

$$\frac{(F_x)_0}{G_0} < f = \tan\phi \qquad (2.14)$$

where G_0 is the submerged weight of the laver A_0:

$$G_0 = \gamma_s n A_0 D_0 = \gamma_s \alpha_2 D_0^3 \qquad (2.15)$$

here n is the "porosity" and $\alpha_2 = n C_2$.

From equation (2.12) and equation (2.14) it is clear that the critical stage of the loose revetment having "small" θ can be expressed as

$$\frac{(F_x)_{0c}}{G_0} = \tan\phi \qquad (2.16)$$

Chapter 2 STABILISATION OF RIVER COURSE

(a) <u>Rip – Rap</u>:

Substituting equation (2.9) and equation (2.13) in equation (2.16) one obtains

$$\frac{\rho U_{0c}^2}{\gamma_s D_0} = (\text{const})_1 \qquad (2.17)$$

(where $(\text{const})_1 = a_1 \tan\phi / C_1$). Since U increases together with the average flow velocity v_0 one can write $U_0 = av_0$ and one can express equation (2.17) in terms of the average flow velocity:

$$\frac{\rho v_0^2}{\gamma_s D} = K \qquad (2.18)$$

(where $K = a_1 \tan\phi / (aC_1)$). Experiment shows that Rip – Rap begins to fall when $K = 2.5$ (Stephenson et al., 1977; Izbash et al., 1970; Naylor et al., 1976) and therefore its critical stage corresponding to "small" θ can be given by

$$\frac{\rho v_0^2}{\gamma_s D_0} \approx 2.5 \quad (\text{Rip – Rap}; \theta \to 0) \qquad (2.19)$$

(b) <u>Rock – Lining</u>:

Substituting equation (2.11) and equation (2.15) in equation (2.16) one obtains

$$\frac{\tau_0}{\gamma_s D_0} = \frac{\rho_0 (v_*)_0^2}{\gamma_s D} (\text{const})_2 \qquad (2.20)$$

(where $(\text{const})_2 = a_2 \tan\phi / C_2$). The left – hand side of equation (2.20) is the familiar Y_{cr}, (which according to the present notation must be denoted by Y_{0c}). As is known from the transport inception curve $(\text{const})_2 \approx 0.05$ and therefore equation (2.20) implies

$$\frac{\tau_0}{\gamma_s D} = \frac{\rho (v_*)_0^2}{\gamma_s D} \approx 0.05 \quad (\text{Rip – lining}; \theta \to 0) \qquad (2.21)$$

[Dividing equation (2.19) by equation (2.21) and taking square root one determines for the friction factor: $c = v_c / (v_{*0c}) \approx 7$. This rather low value of c is reasonable; for in the case of a loose revetment the relative roughness $k_s / h \sim D / h$ is rather large.]

2.2.3 Critical Stage of Loose Revetment for Finite θ

Consider Fig. 2.7. Here the "point" P represents an element of the loose revetment: a stone in the case of Rip-Rap; a rock-layer area $A_\theta = C_2 D_\theta^2$ in the case of rock lining. The bank is inclined by a finite angle θ. The direction of flow, and thus the direction of the force $(\vec{F_x})_\theta$ which is acting on the element P (and which is lying in the plane of the bank) deviate from the horizontal direction by an angle α. The element P will not be dislodged as long as the magnitude of the resultant $\vec{A}$ of the forces $(\vec{F_x})_\theta$ and $\vec{G_\theta'}$ does not exceed the magnitude of the friction force $\vec{R_\theta}$, i.e. as long as $A < R_\theta$ is valid (here A and R_θ are the magnitudes of $\vec{A}$ and $\vec{R_\theta}$).

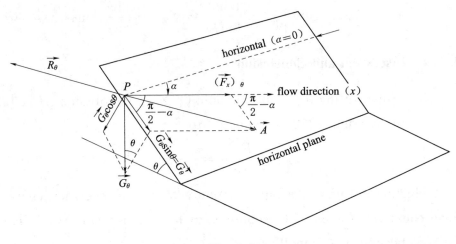

Fig. 2.7

Using "Cosine theorem" one determines for A

$$A = \sqrt{(\vec{F_x})_\theta^2 + (\vec{G_\theta'})^2 + 2\sin\alpha (\vec{F_x})_\theta \vec{G_\theta'}} \qquad (2.22)$$

On the other hand, the magnitude of $\vec{R_\theta}$ is given by

$$R_\theta = (\vec{G_\theta}\cos\theta) f = \vec{G_\theta}\cos\theta \tan\varphi \qquad (2.23)$$

The critical stage occurs when $(\vec{F_x})_\theta$ reaches such a (critical)

Chapter 2   STABILISATION OF RIVER COURSE

value $(\vec{F_x})_{\theta c}$ which yields $A = R_\theta$, i. e. when

$$(\vec{F_x})_\theta^2 + (\vec{G_\theta'})^2 + 2\sin\alpha (\vec{F_x})_{\theta c}\vec{G_\theta'} = (\vec{G_\theta}\cos\theta\tan\phi)^2 \quad (2.24)$$

Substituting $\vec{G_\theta'} = \vec{G_\theta}\sin\theta$, dividing by $\vec{G_\theta}$, and introducing the abbreviation

$$\frac{(\vec{F_x})_{\theta c}}{\vec{G_\theta}} = y_{\theta c} \quad (2.25)$$

one obtains from (2.24)

$$y_{\theta c}^2 + (2\sin\theta\sin\alpha)y_{\theta c} + (\sin^2\theta - \cos^2\theta\tan^2\phi) = 0 \quad (2.26)$$

and solving this second-degree equation one arrives at

$$y_{\theta c} = \frac{(\vec{F_x})_{\theta c}}{\vec{G_\theta}} = \tan\phi\,\psi(\alpha,\theta,\phi) \quad (2.27)$$

where the "function of three angles" $\psi(\alpha, \theta, \phi)$ is given by

$$\psi(\alpha,\theta,\phi) = \cos\theta\left[\sqrt{1 - \left(\frac{\tan\theta}{\tan\phi}\right)^2 \cos^2\alpha} - \left(\frac{\tan\theta}{\tan\phi}\right)\sin\alpha\right] \quad (2.28)$$

2.2.4   Discussion and Conclusions

(1) Substituting $\theta \to 0$ in equation (2.28) one obtains $\psi(\alpha, 0, \phi) = 1$ and equation (2.27) reduces into

$$y_{0c} = \frac{(\vec{F_x})_{0c}}{\vec{G_0}} = \tan\phi \quad (2.29)$$

which is nothing else but equation (2.16). Hence the general form equation (2.28) corresponding to finite θ remains valid also for the special case: "small" θ.

(2) Eliminating $\tan\phi$ from equation (2.27) and equation (2.29) one obtains

$$\frac{(\vec{F_x})_{\theta c}/\vec{G_\theta}}{(\vec{F_x})_{0c}/\vec{G_0}} = \psi(\alpha,\theta,\phi) \quad (2.30)$$

Evaluating the left-hand side of equation (2.30) one determines the following relations:

(a) <u>Rip-Rap</u>:

2.2 Loose Revetment

$$\frac{v_{c\theta}^2 D_0}{v_{c0}^2 D_\theta} = \psi(\alpha,\theta,\phi) \qquad (2.31)$$

(b) Rock – Lining:

$$\frac{\tau_{c\theta} D_0}{\tau_{c0} D_\theta} = \psi(\alpha,\theta,\phi) \qquad (2.32)$$

(3) Now if the flows past the boundaries having finite and "small" values of θ are identical, i.e. if $v_{c\theta} = v_{c0}$ and $\tau_{c\theta} = \tau_{c0}$, then equation (2.31) and equation (2.32) give the same result:

$$D_\theta = \frac{D_0}{\psi(\alpha,\theta,\phi)} \qquad (2.33)$$

Hence the following procedure can be adopted for determining D_θ (required to protect a surface having finite inclination θ from the eroding action of a given flow):

(a) Imagine that the given flow is acting on a bed having "small" θ, and determine first the diameter D_0 (by means of equation (2.19) or equation (2.21)).

(b) Divide D_0 by $\psi(\alpha, \theta, \phi)$ (as implied by equation (2.33)) and obtain D_θ (the value of $\psi(\alpha, \theta, \phi)$ must be computed from equation (2.28)). The values of ϕ corresponding to various materials are given in Fig. 2.8.

(4) Consider now the evaluation of v_{0c} and τ_{0c} which appear in equation (2.19) and equation (2.21). The loose revetment must remain intact even when the river flow is in its "strongest" state, and therefore v_{0c} and τ_{0c} must be determined from the conditions corresponding to the largest flood flow rate Q_f (that is likely to occur during the period of time for which the revetment is designed). Usually, the flow rate of a certain "M – year flood" is adopted for this purpose.

(a) Straight River Region: In this case $\alpha = 0$ and equation (2.28) reduces into

Chapter 2 STABILISATION OF RIVER COURSE

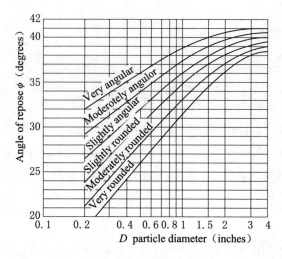

Fig. 2.8 Angles of Repose of Noncohesire Material after U. S Bureau of Reclamation

$$\psi(\alpha,\theta,\phi)=\cos\theta\sqrt{1-\left(\frac{\tan\theta}{\tan\phi}\right)^2} \qquad (2.34)$$

1) Rip-Rap: The part of Q_f flows over the flood plain and we have an "overtopping" Δh (Fig. 2.9).

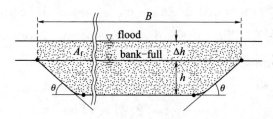

Fig. 2.9

Consider the flood average velocity v_f in the "effective area" A_f. Write Manning formula for "bank-full" and "flood" stages respectively:

$$v=\frac{1}{n}R^{\frac{2}{3}}S^{\frac{1}{2}} \quad \text{and} \quad v_f=\frac{1}{n}R_f^{\frac{2}{3}}S_f^{\frac{1}{2}} \qquad (2.35)$$

Dividing these expressions, bearing in mind that wetted perimeters for A and A_f are the same, and taking into account that

2.2 Loose Revetment

(owing to the usually "large" value of the width/depth ratio) $A/B = h$ and $A_f/B = h + \Delta h$, one determines

$$v_f = v \left(\frac{S_f}{S}\right)^{\frac{1}{2}} \left[1 + \frac{\Delta h}{h}\right]^{\frac{2}{3}} \quad (2.36)$$

where

$$v = \frac{Q_{bf}}{A} \quad \text{and} \quad S \approx \text{river bed slope} \quad (2.37)$$

When using equation (2.19) the velocity v_{0c} must be identified with v_f. If the value of v_f is not known (from field or model measurements) then it can be estimated with the aid of equation (2.36). Here the slope S_f must be interpreted as a "free surface slope of the flood waver" while Δh as the corresponding "overtopping". Although the maximum values of S_f and Δh do not occur simultaneously, it would be safer and simpler (and yet the "safety" will not be exaggerated much) if equation (2.36) is evaluated simply by using the largest values of S_f and Δh which occur during the adopted "M - year flood". Substituting $v_{0c} = v_f$ in equation (2.19) one determines first D_0 (as if to protect the bed). Subsequently one determines D_θ (from equation (2.33)) which is actually needed to protect the bank.

2) Rock - Lining: The largest shear stress corresponding to a given flow is acting at the lowest part of the bank (points b and c in Fig. 2.9). The largest value of this largest stress is when the product (hS) is maximum. Thus one can adopt

$$\tau_{0c} = \gamma h_f S_f = \gamma (h + \Delta h) S_f \quad (2.38)$$

where Δh and S_f are maximum values of the "overtopping" and the flood wave slope (although, as has been already mentioned, they do not occur simultaneously).

(b) River Bend: In this case, the river cross - section is not symmetrical (as in the case of an "ideal river" in a straight region);

Chapter 2 STABILISATION OF RIVER COURSE

it has a shape as shown schematically in Fig. 2.10a (Rosovskii et al., 1957; Odgaard et al., 1982; Grishin et al., 1955; etc). The flow is helicoidally: the forward-moving fluid mass is rotating with downwards directed velocities at the outer bank (which may require protection). Thus a is not "zero": it is of the order of $\alpha = 10°$ to 15° (Garde et al., 1977; Rosovskii et al., 1957; Odgaard et al., 1982), and the "function of three angles" $\psi(\alpha, \theta, \phi)$ must be evaluated by using its general form equation (2.28).

Note, from Fig. 2.10a that the largest velocities at the cross-section of a bend are (approximately) above the deepest point b of the cross-section and near the outer bank. The fact that u_{max} is (approximately) above the deepest point b means that the velocity field in the sub-area cab (contacting the outer bank) should be approximately the same as that of the right half of the hypothetical symmetrical cross-section $a'ab$ conveying the same flow rate Q (Fig. 2.10b); the sub-area cbd of the actual cross-section being then but a "stretched-to-the-left-version" of the left half cba' of the hypothetical cross-section. Let A be the area of the actual cross-section abd and A_S of the (reduced) symmetrical cross section $a'ab$. According to the explanations above the flow velocity acting on the outer bank should be Q/A_S (rather than Q/A). It follows that when dealing with the Rip-Rap at the bend one should increase the velocity $v_{0c} = v_f$ (given by equation (2.36) by multiplying it with the (larger than unity)) "bend factor":

$$\alpha_r = \frac{A}{A_s} \quad (2.39)$$

Note from Fig. 2.10c that the elevation of the lowest part b' of the loose revetment at the outer bank should be comparable with that of the deepest point b of the cross-section, and therefore when dealing with the rock—lining at the bend the value of τ_{0c} must be de-

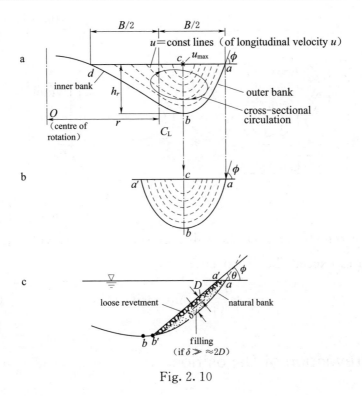

Fig. 2.10

termined (from equation (2.38)) by using the largest bank full flow depth h_r of the cross-section.

(5) The slope S of a meandering river at the bank full stage is comparable with the average slope of the river channel bed. Hence the slope S is, as a rule, smaller than the slope S_v of the river valley (for, as shown in Fig. 2.11, the length l_{AB} of the river channel between the sections A and B is larger than the direct valley distance $\overline{AB}$).

During the flood, when banks are overtopped by an amount Δh, the fluid near the free surface is moving in the direction of the valley (broken arrows) and the slope of the free surface is comparable with the valley slope S_v (rather than with the slope of the river channel), the largest free surface slope S_f of the flood wave being larger than S_v. Hence, when using equation (2.36), it may be

Chapter 2 STABILISATION OF RIVER COURSE

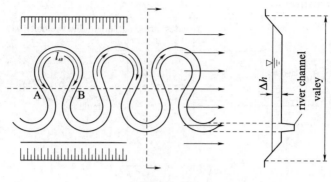

Fig. 2.11

helpful to remember that the value of the ratio (S_f/S) should be expected to exceed the "river sinuosity":

$$\frac{S_f}{S} > \frac{S_v}{S} = \frac{l_{AB}}{AB} = \sin(uosity)(\geqslant 1) \qquad (2.40)$$

2.3 Prevention of Deposition

Deposition and thus the increment of the elevation z of a river bed takes place mostly in the lower reaches (Zone III in Fig. 1.3) and sometimes in the middle reaches (Zone II). In subsection 1.2 it has been explained that Zone III is, in general, a zone of deposition (where $\partial z/\partial t > 0$). However, these positive values of $\partial z/\partial t$, which characterize the time increment of the elevations of the river bed as a whole must be considered as "very small". Similarly, in Zone II the overall value of $\partial z/\partial t$ is "zero". This, however, does not mean that in some regions of these zones $\partial z/\partial t$ can't, acquire some values that are much larger than what is meant by "very small" and "zero". These will be the regions of "local depositions". Consider for example a river section A where the slope S_1 suddenly changes into S_2 (Fig. 2.12a). Since $S_1 > S_2$, we have $\tau_{01} > \tau_{02}$ and thus $q_{S_1} > q_{S_2}$. But this means that the sediment entering

2.3 Prevention of Deposition

section A cannot be transported by the river further downstream and the occurrence of a local deposition along a certain region $\overline{AB}$ must be expected. In other words a larger than average $\partial q_s/\partial x \sim (q_{s_2}-q_{s_1})/\overline{AB}$ gives rise to the larger than average values of $\partial z/\partial t$ (eqn (1.4)). Similarly, a tributary with a steep slope $S_1 (\gg S)$, and thus a high erosion capacity, can also bring (continually "inject") into the river such an amount of sediment that the river will not be able to transport it further totally. This will also create a local deposition (Fig. 2.12b).

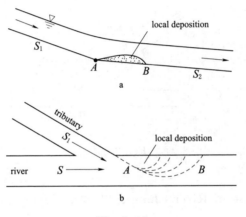

Fig. 2.12

The depositions caused by the factors described above, or to that matter by many other factors, may be regarded as undesirable and one may intend to eliminate or prevent them. This can be achieved by the following methods.

2.3.1 Reduction of the Incoming Sediment

Reduce the amount of the "injected" sediment by reducing the erosion capacity of tributaries and/or of the river itself fortify the side slopes of the valley in the upper reaches (Fig. 2.13a), build "sediment traps" (Fig. 2.13b), remove the sediment of a (much sediment carrying) tributary by diverting it first into a deep lake

Chapter 2   STABILISATION OF RIVER COURSE

which can act as a "settling basin" (Fig. 2.13c), if such a lake exists in the area, etc (Duhm et al., 1951; Press et al., 1956; Schaffernak et al., 1950; Grishin et al., 1955).

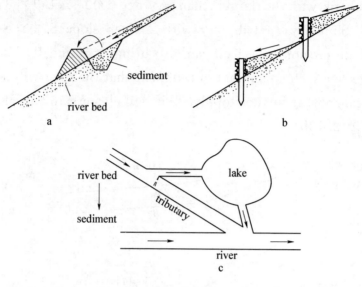

Fig. 2. 13

2.3.2   Reduction of River Channel Width; Dykes and Groynes

This method is the opposite to that described in 2.1.1. The intention here is to reduce the river width (from B_1 to B_2) and thereby to achieve the increment of the flow depth (from h_1 to h_2). The increment of h will lead to the increment of $\tau_0 \sim h$: consequently, it will lead to the increment of the transport rate and thus to the reduction of deposition (Duhm et al., 1951; Grishin et al., 1955). Using the same reasoning as that leading to equation (2.5) one can obtain for the present case

$$\frac{h_1}{h_2} = \left(\frac{B_2}{B_1}\right)^{\frac{3}{5}} (<1) \tag{2.41}$$

The reduction of the river width (from B_1 to B_2) is carried out either with the aid of dykes (Fig. 2.14a) or with the aid of groynes

(Fig. 2.14b). Very often the combination of both is utilized (Fig. 2.14c). If one of the banks is "strong", then dykes or groynes can be constructed only at the "weak" bank (it is assumed that the "strong" bank will withstand the action of the increased τ_0).

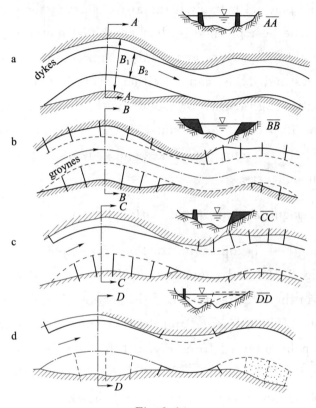

Fig. 2.14

As a rule, groynes are more economic than dykes. Furthermore, with the aid of groynes the river width can be reduced gradually and this is a very important advantage. Indeed the exact value of the reduced width B_2 cannot be known beforehand. Yet by building the groynes (starting from the banks towards the middle of the stream) one can terminate the construction process precisely at the stage which will be observed to be as the "most appropriate". This kind of "experiment – in – the – river – itself" cannot be carried out when

Chapter 2 STABILISATION OF RIVER COURSE

building the dykes along the "predetermined" lines. On the other hand, the head of groynes is prone to erosion, the flow past them is not as smooth and regular as that past the dykes, while the deposition (which is expected to fill the spaces between the groynes) often does not take place effectively if the groynes are at the outer bank. It follows that the best method is the "combined method": whenever possible, build the groynes at the inner bank and dykes at the outer bank. Groynes and dykes can be either "transparent" (with some flow passing through them) or "solid" (with no flow passing through them); an idea on their construction can be gained from Fig. 2.15. The arrangement of dykes and groynes in plane is best revealed by model tests.

2.3.3 Increment of Slope ("cut_offs")

This method is opposite to that presented in 2.1.1 (1). Consider Fig. 2.16a and suppose that the (undesirable) deposition takes place in the river region $\overline{ab}$. If the sections b and d are connected by a "cut-off" channel $\overline{bd}$ and the flow is diverted from $\widehat{bcd}$ to $\overline{bd}$ then (as can be inferred from Fig. 2.16b) the existing river slope $S_1 = \Delta H / \widehat{bcd}$ will increase to $S_2 = \Delta H / \overline{bd}$. This increment of slope will lead to the increment of $\tau_0 \sim S$, and thus q_s, and consequently to the reduction of deposition along $\overline{ab}$. In fact the execution of a "cut-off" may lead even to the erosion at b. Indeed with the passage of time the river tends to "iron out" the artificial "corners" at b and d (by eroding the bed at b and by depositing the material at d as shown schematically by times t_0, t_1, t_2... in Fig. 2.16c). Hence this method must be used with some caution (in order not to "over do" when preventing the erosion) and the more so the larger is the ratio $\widehat{bcd}/\overline{bd}$.

Since the slope S_2 of a cut-off channel is larger than the river

2.3 Prevention of Deposition

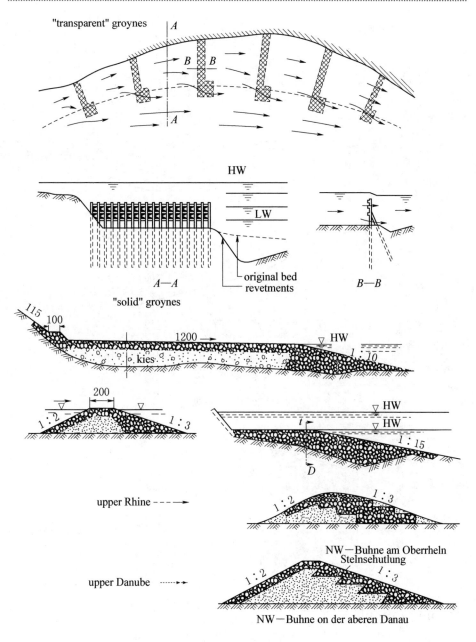

Fig. 2.15

slope S_1, the cut-off channel can convey the flow rate Q by means of a channel width B_2 and the flow depth h_2 that are smaller than their river counterparts B_1 and h_1, say, in that region. Writing the

Chapter 2   STABILISATION OF RIVER COURSE

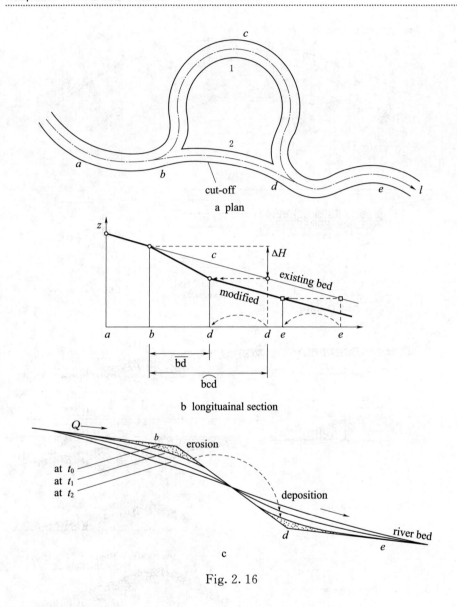

Fig. 2.16

Manning formula

$$Q = Bh \frac{k}{n} h^{\frac{2}{3}} S^{\frac{1}{2}} \qquad (2.42)$$

for river and cut-off channels (i.e. for subscripts 1 and 2) and equating them we obtain

2.3 Prevention of Deposition

$$\frac{B_1 h_1^{\frac{5}{3}} S_1^{\frac{1}{2}}}{n_1} = \frac{B_2 h_2^{\frac{5}{3}} S_2^{\frac{1}{2}}}{n_2} \quad (2.43)$$

And thus

$$\frac{B_2}{B_1} = \left(\frac{h_1}{h_2}\right)^{\frac{5}{3}} \left(\frac{S_1}{S_2}\right)^{\frac{1}{2}} \left(\frac{n_1}{n_2}\right) \quad (2.44)$$

Knowing the river channel characteristics B_1, h_1 and n_1 can determine any one of the three cut-off characteristics B_2, h_2, and n_2 (by adopting the remaining two) from the equation (2.44).

Chapter 3 MODIFICATION OF RIVER COURSE

3.1 General

Suppose, that the existing irregular river channels I in Fig. 3.1a and Fig. 3.1b will be modified into regular channels II. What must be the width, depth, and slope (B, h, and S) of the new channels II? Can the designer choose for B, h, and S any values that he may think appropriate? The answer to this relevant question can be understood better if the, so-called, "Regime concept" is introduced first.

3.1.1 Regime Concept

Imagine an unbounded erodible granular medium with a plane surface, and suppose that we have excavated in it a trapezoidal channel having free surface width B_0, depth h_0, and slope S_0 (Fig. 3.2a). Let us now assume that a constant flow rate Q is admitted into this channel and that this Q will flow in the channel forever. What will happen to our erodible (and thus deformable) channel?

Such a hypothetical experiment cannot be carried out in practice, and "what will happen to the channel" can only be guessed from observations and measurements performed under the resembling conditions. The prevailing contemporary view is that, in general, the flow rate Q will not "accept" our channel, but that it will deform it

3.1 General

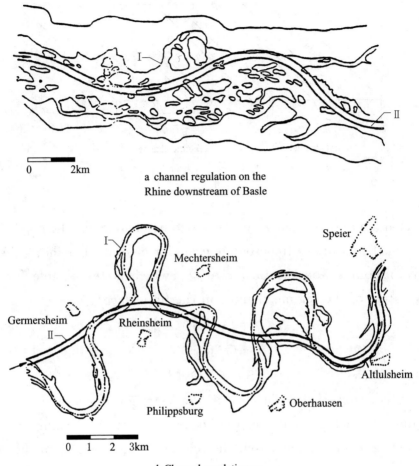

a channel regulation on the
Rhine downstream of Basle

b Channel regulation on
the Rhine upstream of Mannheim

Fig. 3.1

as to establish, eventually, a certain definite channel "of its own" (Fig. 3.2b) which is referred to as the "regime channel " (of that Q). The experiment suggests that once the regime channel is established, it will not change with time any longer, or that the regime channel is stable. It follows that the characteristics $\overline{B}$, $\overline{h}$ and $\overline{S}$ of the established regime channel should not be the functions of time and furthermore they should have nothing in common which the characteristics B_0, h_0 and S_0 of the "initial channel" . But this means that

Chapter 3　MODIFICATION OF RIVER COURSE

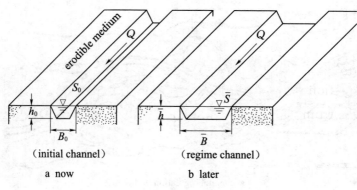

Fig. 3.2

the characteristics $\overline{B}$, $\overline{h}$ and $\overline{S}$ of a regime channel can be dependent only on the nature of the erodible granular medium (in short: "M") and, of course, on the flow rate Q (which is responsible for the generation of the regime channel in the first place):

$$\left.\begin{aligned} \overline{B} &= f_B(Q,M) \\ \overline{h} &= f_h(Q,M) \\ g\overline{S} &= f_S(Q,M) \end{aligned}\right\} \quad (3.1)$$

It is assumed that the flow in the (straight and infinitely long) regime channel is uniform (or quasi-uniform (Yalin et al., 1977)) and consequently that the role of S consists only of the generation of the component $\vec{g}S$ acting per unit fluid mass in the flow direction. Since the component $\vec{g}\cos\theta = \vec{g}$ (perpendicular to the flow direction) produces only the hydrostatic pressure which has no bearing on the progress of the present phenomenon, the consideration of the product $g\overline{S}$ eliminates any further necessity of the parameters g and $\overline{S}$ in the formulation of the regime phenomenon.

　　Consider now the formation of a river channel in a (reasonably) homogeneous alluvium. This channel is formed by the river flow itself and therefore it is a regime channel. However, in contrast to the regime channel explained above, the river regime channel has not been formed by a constant Q, but by a variable Q (by $Q =$

$\phi_Q(t)$ varying within a certain interval $Q_{min} < Q < Q_{max}$). It is assumed nowadays that the characteristics $\overline{B}$, $\overline{h}$ and $\overline{S}$ of the river regime channel (formed by a variable flow rate) are also determined by the equation (3.1) provided that instead of Q an appropriate flow rate (from the interval $Q_{min} < Q < Q_{max}$) is substituted in them. The view of the majority of researchers is that this "appropriate" flow rate must be the bankfull flow rate Q_{bf} (Leopold et al., 1964; Garde et al., 1977; Henderson et al., 1966; etc).

3.1.2 Modified River Channel

Let us now turn again to the question of width, depth, and slope of the modified river channel. From the explanations above it should be clear that the modified river channel II (Fig. 3.1) can then be expected to remain stable (i.e. not deform and/or displace in time) if its width, depth, and slope (B, h and S) are assigned the river regime values $\overline{B}$, $\overline{h}$ and $\overline{S}$ which are determined by equation (3.1) where $Q = Q_{bf}$ (Garde et al., 1977; Henderson et al., 1966; Blench et al., 1966; etc).

$$\left. \begin{aligned} B = \overline{B} = f_b(Q_{bf}, M) \\ h = \overline{h} = f_h(Q_{bf}, M) \\ gS = g\overline{S} = f_s(Q_{bf}, M) \end{aligned} \right\} \quad (3.2)$$

But why should one compute the values of $\overline{B}$, $\overline{h}$ and $\overline{S}$ (from the "regime equation" (3.2)) and not simply measure them in the river itself? There is no need indeed if the river is regular and $\overline{B}$, $\overline{h}$ and $\overline{S}$ are measurable. The point is that it is often a river which has an irregular geometry that needs the improvement (such as I in Fig. 3.1) and in this case, it may be rather difficult to decide as to what the values of $\overline{B}$, $\overline{h}$ and $\overline{S}$ might be (and consequently what B, h and S of the planned channel II should be). Furthermore, what if the river is braiding and we intend to "canalize" it (i.e. to make it flow in a single channel)? In this case, a single set ($\overline{B}$, $\overline{h}$ and $\overline{S}$) may not be present at all.

3.2 Regime Channels

Regime channels have attracted the attention of many researchers, and therefore a very large number of "regime equations" have been suggested to date. Yet the problem of regime channels is far from being solved. A convincing theory for their formation has not been worked out, the regime equations proposed are purely empirical, and most of them are dimensionally inhomogeneous (which does not only mean that they are scientifically unsound, but that they cannot be used as they stand in every system of units). Hence there is no point in reviewing all of these equations and it will be confined here to the presentation of only two sets of regime equations (which can be regarded as the most prominent ones).

3.2.1 Regime Equations of T. Blench (Blench et al., 1966; Blench et al., 1957; Henderson et al., 1966; Garde et al., 1977)

According to Blench, the equation (3.1) can be expressed as follow

$$\left. \begin{array}{l} \overline{B} = \alpha_B Q^{\frac{1}{2}} \\ \overline{h} = \alpha_h Q^{\frac{1}{3}} \\ \overline{S} = \alpha_S Q^{-\frac{1}{6}} \end{array} \right\} \quad (3.3)$$

These equations must be used in the British system of units only: lengths in feet, times in seconds. The dimensional coefficients α_B, α_h and α_S are determined by

$$\alpha_B = \sqrt{\frac{f_0}{f_\theta}}, \; \alpha_h = \sqrt[3]{\frac{f_\theta}{f_0^2}}, \; \alpha_S = \frac{f_0^{\frac{5}{6}} f_\theta^{\frac{1}{12}} v^{\frac{1}{4}}}{3.63 g (1 + 430 c_0)} \quad (3.4)$$

where f_0 is the "bed factor"
$$f_0 = 33.2\sqrt{D}(1+12000C_0) \tag{3.5}$$
while f_θ is the "bank factor"
$$\left.\begin{aligned} f_\theta &= 0.1 \text{(slightly cohesive banks)} \\ f_\theta &= 0.2 \text{(moderately cohesive banks)} \\ f_\theta &= 0.3 \text{(highly cohesive banks)} \end{aligned}\right\} \tag{3.6}$$

In the expressions above, D is the typical size (in feet) of the flow boundary material, C_0 is the dimensionless volumetric concentration at the bed, v kinematic viscosity (ft²/s), and g acceleration due to gravity ($=32.2$ft/s²).

3.2.2 Regime Equations of D. B. Simons and M. L. Albertson (Hendersonet al., 1966; Simonset al., 1960; Raudkiviet al., 1976)

According to these Authors
$$\left.\begin{aligned} \overline{B} &= 0.9K_1 Q^{\frac{1}{2}} \\ \overline{R} &= K_2^{0.36} \quad \text{(hydraulic radius)} \\ \overline{h} &= 1.21\overline{R} \quad \text{(if } R>7\text{ft)} \\ \overline{h} &= 2+0.93\overline{R} \quad \text{(if } R>7\text{ft)} \\ \overline{v} &= Q/(\overline{B}\overline{h}) = K_3(\overline{R}^2 S)^m \end{aligned}\right\} \tag{3.7}$$

These equations are also intended for "feet" and "second" units, and in this case, we have for K_i and m

	K_1	K_2	K_3	m
(1) sand bed and sand banks	3.50	0.52	13.30	0.33
(2) sand bed, cohesive banks	2.60	0.44	16.00	0.33
(3) cohesive bed, cohesive banks	2.20	0.37	17.00	0.31
(4) coarse non-cohesive bed and banks	1.75	0.23	17.90	0.29
(5) same as (2) but "large" C_0 (>0.02)	1.70	0.34	16.00	0.29

However, one can use equation (3.7) also in metric units (all lengths in meters, Q in m³/s) provided that K_i and m are taken

Chapter 3 MODIFICATION OF RIVER COURSE

from the table below (Raudkivi et al., 1976)

	K_1	K_2	K_3	m
(1) sand bed and sand banks	6.34	0.57	9.28	0.33
(2) sand bed, cohesive banks	4.71	0.48	10.67	0.33
(3) cohesive bed, cohesive banks	3.98	0.41	10.75	0.31
(4) coarse non-cohesive bed and banks	3.17	0.25	10.87	0.29
(5) same as (2) but "large" C_0 (>0.02)	3.08	0.37	9.71	0.29

When equation (3.3) to equation (3.7) are used for the design of a (modified) river channel then it would be only reasonable to use as Q the bankfull flow rate Q_{bf} of the existing river channel (in that region). One would expect that the computed values of $\overline{B}$ and $\overline{h}$ will approximate those which will be present in the designed channel at the bankfull stage.

3.2.3 Banks of a Regime Channel

A regime channel, formed under ideal conditions, has the shape of a curvilinear trapezoid (Fig. 3.3a) and it is intended now to reveal the shape of its curvilinear banks. Consider Fig. 3.3b.

The value of the shear stress τ_0 acting at a point M of the bank $\widehat{oa}$ is given by

$$\tau_\theta \delta S = \gamma \overline{B} \delta A \qquad (3.8)$$

and since

$$\delta A = \delta A_y = y \delta S \cos\theta \qquad (3.9)$$

one obtains

$$\tau_\theta = \gamma \overline{S} y \cos\theta \qquad (3.10)$$

which gives for the (deepest) point a

$$\tau_\theta = \gamma \overline{S} h \qquad (3.11)$$

Hence

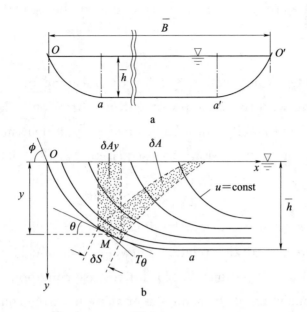

Fig. 3.3

$$\frac{\tau_\theta}{\tau_0} = \frac{y}{h}\cos\theta \tag{3.12}$$

It can be shown in a rigorous manner (with the aid of the Exner Polya equation) that if the "widening" of an alluvial channel is terminated, (i. e. if its regime state ($\overline{B}$, $\overline{h}$, $\overline{S}$) is achieved and its banks remain constant in time) then the transport of bank material is also terminated ($q_s \equiv 0$ at all points M having $\theta \geqslant 0$) and the shear stress τ_θ acting at a point M of the bank is reduced to its respective critical value $\tau_{c\theta}$:

$$\tau_\theta = \tau_{c\theta} \tag{3.13}$$

Accordingly, the relation (3.12) must be interpreted as follows

$$\frac{\tau_{c\theta}}{\tau_{c\theta}} = \frac{y}{h}\cos\theta \tag{3.14}$$

It is assumed that the regime channel under consideration is either straight or that its curvature in plane is so "small" that the in-

Chapter 3 MODIFICATION OF RIVER COURSE

fluence of the cross-sectional circulations at the bends can be neglected and

$$\alpha \approx 0 \qquad (3.15)$$

can be adopted (see section 3.2 for the meaning of the angle α). When dealing with the loose revetment in subsection 3.2.3 we have found that if the stability of a bank formed by a loose material isendangered by the shear action, then equation (2.18), i.e.

$$\frac{\tau_{c\theta} D_0}{\tau_{c0} D_\theta} = \psi(\alpha, \theta, \phi) \qquad (3.16)$$

is valid. Subsequently, using this formula for $\tau_{c\theta} = \tau_{c0}$ and $\theta = $ const we have determined the relation between (nonequal) D_θ and D_0. Here we will use equation (3.16) for reverse conditions. In the case of a bank consisting of the same material we have the constant diameter throughout:

$$D_\theta = D_0 = D \qquad (3.17)$$

On the other hand, the shear stresses and angles θ are not equal: they vary (from one point M to another) as functions of y. Substituting equation (3.14) and equation (3.17) in equation (3.16) and taking into account that

$$\tan\theta = \frac{dy}{dx} \qquad (3.18)$$

we obtain (with the aid of equation (2.34) which corresponds to $\alpha = 0$)

$$\frac{y}{x} = \sqrt{1 - \frac{1}{\tan^2\phi}\left(\frac{dy}{dx}\right)^2} \qquad (3.19)$$

and thus

$$\frac{dy}{dx} = \tan\phi \sqrt{1 - \left(\frac{y}{h}\right)^2} \qquad (3.20)$$

which is the differential equation of the contour $\widehat{oa}$ of the bank. Integrating equation (3.20) one obtains

$$\frac{y}{h} = \sin\left(\tan\phi \, \frac{x}{h}\right) \tag{3.21}$$

Hence the bank $\overset{\frown}{oa}$ of a (nearly) straight regime channel has the form of a sine curve.

3.2.4 "Stable" Channels, "Live Bed" Channels

(1) The contemporary convention is to assume that the flow over the (horizontal) channel bed $\overline{aa}'$ (see Fig. 3.3a) is two-dimensional (Henderson et al., 1966; Chow et al., 1959; Graf et al., 1971; etc).

This is a sound assumption as long as D is sufficiently large (larger than 1cm or so). On the other hand, if D is so large then it is unlikely that such a channel (which, from a practical standpoint, is perfectly conceivable) is an alluvial regime channel; for alluvium usually is sand – not gravel. Still, let us comply with the existing convention and let us thus assume that the flow over $\overline{aa}'$ is two-dimensional (Fig. 3.4a).

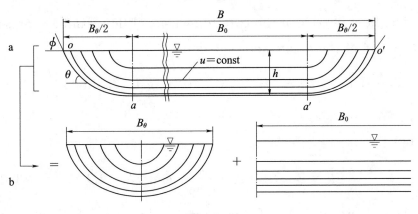

Fig. 3.4

This means that it is assumed that all points of the channel bed $\overline{aa}'$ are subjected to the same conditions and therefore if the shear stress acting at the point a (which is the lower limit of the bank) is

Chapter 3   MODIFICATION OF RIVER COURSE

"critical" (i.e. it is τ_{0c} as explained in the text leading to equation (3.13)) then it must also be critical at any other point of the channel bed $\overline{aa}'$. But this means that all points of the perimeter $oaa'o'$ (Fig. 3.4a) are under the action of critical shear stresses, and consequently that all grains forming the flow boundary $oaa'o'$ are in the "critical state" (just below the "threshold" of movement). A channel like this is called "stable channel".

Let us now consider the conveyance capacity of a stable channel. Since the flow above $\overline{aa}'$ is assumed to be two-dimensional its isovels ($u = $ const lines) must be regarded as horizontal straight lines, and the isovels of the regions above the banks must merge into them tangentially. But if so then the total flow can be split into two components (Fig. 3.4b): one flowing in a two-dimensional channel (which has the width B_0 and which will be marked by the subscript 0). And the other flowing in the sinusoidal channel (which will be marked by the subscript θ). i.e.

$$Q = Q_0 + Q_\theta \tag{3.22}$$

The flow rate Q_0 of the two-dimensional flow can be expressed with the aid of the logarithmic relation

$$Q_0 = B_0 q_0 = B_0 h v = B_0 h v_* c_0 = B_0 h \sqrt{gsh} \left[2.5 \ln \left(11 \frac{h}{k_{S_0}} \right) \right] \tag{3.23}$$

where

$$B_0 = B - B_\theta = B - \frac{\pi}{\tan\phi} h \text{ and } k_{S_0} = 2D \tag{3.24}$$

The flow rate Q_θ of the three-dimensional flow in the sinusoidal channel can be given by the Manning formula

$$Q_\theta = A_\theta v_\theta = \frac{K}{n_\theta} S^{\frac{1}{2}} A_\theta K_\theta^{\frac{2}{3}} \tag{3.25}$$

which using

3.2 Regime Channels

$$A_\theta R^{\frac{2}{3}} = \frac{2h^2}{\tan\phi}(0.57h)^{\frac{2}{3}} = \frac{1.357}{\tan\phi}h^{\frac{8}{3}} \qquad (3.26)$$

and

$$\frac{K}{n_\theta} = \frac{8.4\sqrt{g}}{K_{S_\theta}^{\frac{1}{6}}} = \frac{8.4\sqrt{g}}{2^{\frac{1}{6}}D^{\frac{1}{6}}} \qquad (3.27)$$

Q_θ can be written as

$$Q_\theta = \frac{8.4\sqrt{g}}{2^{\frac{1}{6}}D^{\frac{1}{6}}} S^{\frac{1}{2}} \frac{1.357}{\tan\phi} h^{\frac{8}{3}} \qquad (3.28)$$

and thus as

$$Q_\theta = \frac{10.3}{\tan\phi}\left[\frac{h}{D}\right]^{\frac{1}{6}} h^2 \sqrt{gSh} \qquad (3.29)$$

Substituting equation (3.23) and equation (3.29) in equation (3.22), and considering equation (3.24) obtains

$$\frac{Q}{\sqrt{gSh}} = h^2 \left\{ \left[\frac{B}{h} - \frac{\pi}{\tan\phi}\right]\left[2.5\ln\left(5.5\frac{h}{D}\right)\right] + \frac{10.3}{\tan\phi}\left[\frac{h}{D}\right]^{\frac{1}{6}} \right\} \qquad (3.30)$$

Furthermore, the fact that the channel is in the critical stage means that h, S, and D are interrelated as

$$\frac{\gamma}{\gamma_s}\frac{Sh}{D} \approx 0.05 = Y_{cr} \qquad (3.31)$$

(2) Let us now return to regime channels (formed in alluvial sand). Observations (in field and laboratory) indicate that the bed of such channels usually is covered by sand waves and thus that some transport q_s must be present on the channel bed $\overline{aa'}$. This fact sometimes is termed as the "Regime - Channel Paradox" (Grafet al., 1971): because "if the flow is critical at all points of the flow boundary $oaa'o'$ then no transport must be present ($q_s \equiv 0$ all over)". In face, there is no paradox here and the misinterpretation is entirely due to the assumption that the flow over $\overline{aa'}$ is two-dimensional. Indeed, consider Fig. 3.5.

There is no physical reason that would compel the flow over the

Chapter 3 MODIFICATION OF RIVER COURSE

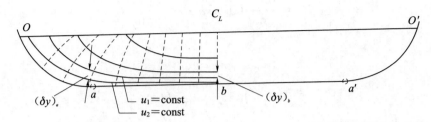

Fig. 3.5

bed aa' to possess the isovells that are horizontal straight lines; the only constraint is that isovels must be symmetrical concerning the centre line (and thus that they must possess horizontal tangents over the point b). Let δy be the distance between isovels u_1 and u_2 ($\delta u = u_2 - u_1$). Since $(\delta y)a > (\delta y)b$ it is clear that

$$\tau_b = \rho l^2 \left[\frac{\delta u}{(\delta y)_b}\right]^2 > \rho l^2 \left[\frac{\delta u}{(\delta y)_a}\right]^2 = \tau_a = \tau_{cr} \qquad (3.32)$$

But if $\tau_b > \tau_{cr}$ then $(q_s)_b > 0$. Admittedly the curvature of isovells immediately above $\overline{aa'}$ is very small; they are almost horizontal straight lines (particularly if the bed roughness (D) is "large"), the assumption that isovells are horizontal straight lines and thus that the flow over $\overline{aa'}$ is two dimensional is therefore "good enough" for the computation of conveyance capacity, however, it is not "good enough" to pass the judgment on the presence or absence of sediment transport. It should be remembered that most of the regime channels in fine alluvium do transport some sediment, and they are termed therefore as the "live bed channels"; the transport rate q_s (per unit flow width) progressively increases (from zero onwards) along with the interval $(a \to b)$.

3.3 Curvilinear River Regions

The Regime equations introduced above correspond to (reason-

3.3 Curvilinear River Regions

ably) straight regime channels. Yet, as has been pointed out, 1.3, the channel of a natural river sometimes is (reasonably) straight indeed, but sometimes it is braiding or meandering: this depends on the river channel slope S.

Let S_* be the slope corresponding to the initiation of meandering. The value of S_* can be determined (or estimated) from the Leopold and Wolman (Leopold et al., 1964) formula

$$S_* = 0.0116 Q^{-0.44} \qquad (3.33)$$

(where $[Q] = m^3/s$), or from the Henderson (Henderson et al., 1966) formula

$$S_* = 0.64 D^{1.14} Q^{-0.44} \qquad (3.34)$$

(where $[Q] = ft^3/s$).

No "initial channel" has been provided for a natural river that would "compel" the river to flow in it, and thus prevent it from braiding (i.e. from flowing in a multitude of channels). If, however, a channel is provided (such as e.g. the man-made channel Ⅱ in Fig. 3.1) then the "freedom" of the river is restricted and its braiding is prevented. The same can be said for the man-made canals and the laboratory regime-test flows (which originate in an "initial channel" B_0, h_0, S_0).

Hence except some seldom cases (e.g. such as those corresponding to comparatively large values of the ratio $\Delta H/h$ (see Fig. 2.9)) a modified (and thus "canalized") river will not be braiding: it will be compelled to flow in a single channel which might be either straight (if $S > S_*$) or meandering (if $S < S_*$).

The straight line in Fig. 3.6 is the graph of equation (3.33). This line separates the zone of "meandering" from the zone of "straight or braiding": all of the points corresponding to "braiding" in this graph are due to natural river courses.

Chapter 3 MODIFICATION OF RIVER COURSE

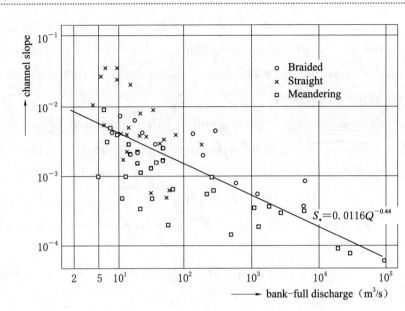

Fig. 3. 6

3.3.1 Meandering

Suppose that the straight regime channel ($\overline{B}$, $\overline{h}$, $\overline{S}$) has been formed (Fig. 3. 7). If $S<S_*$ then the flow in it is not stable: the slightest "geometric discontinuity" (G in Fig. 3. 7b) will deform the straight (rectilinear) pattern of the flow into a sinusoidal one (in plane). The stream lines of the sinusoidal flow are "tightly packed" at the locations E and they are "loosely packed" at D. Hence the velocities and shear stresses at E are larger while those at D are smaller than the critical velocities and shear stresses on the banks of the original rectilinear flow (in Fig. 3. 7a). This will inevitably lead to the initiation of erosion at E and of the deposition at D. Consequently it will lead to a sinusoidal deformation of the channel boundaries in plane (Fig. 3. 7c): the channel will begin to meander. The amplitude at of the meandering flow will grow (as shown schematically in Fig. 3. 7d) until such a large loop length $\widehat{ab}$

3.3 Curvilinear River Regions

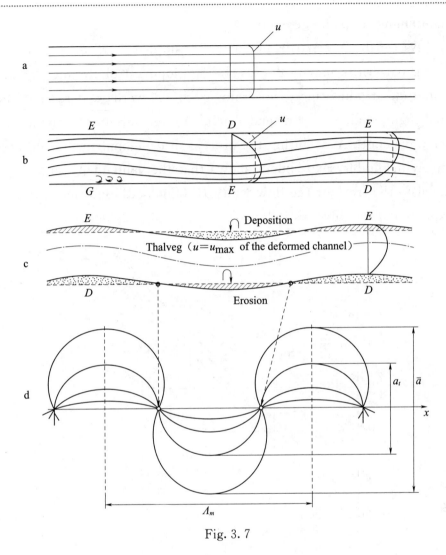

Fig. 3.7

and thus such a small slope $S \sim 1/\widehat{ab}$, is reached that the velocities at E (which decrease in proportion to $\sqrt{S}$) will not be able to erode this (concave) bank further. At this stage, the growth of the meandering amplitude will terminate, and the final (regime) geometry of the meandering stream will be achieved. (This, however, does not mean that the meandering river channel will necessarily be static. Indeed, it may "migrate" in the direction x (like a snake)

Chapter 3 MODIFICATION OF RIVER COURSE

maintaining its plane geometry.)

As can be inferred from Fig. 3.7d the development of a meandering pattern is mainly due to the development of the meandering amplitude at which increases from "zero" onwards and reaches the final regime value $\bar{a}$ ("meander belt"): the variation of the meander wavelength Λ_m, during the process of development, is comparatively negligible. Experiment shows that the value of Λ_m is proportional to the free surface flow width B (Fig. 3.8) and that it can be expressed (roughly) as follows:

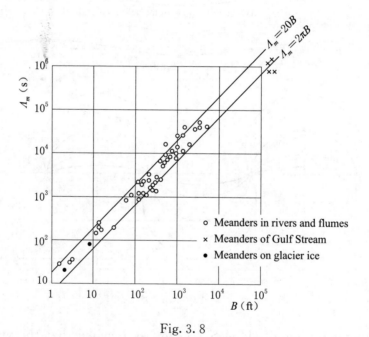

Fig. 3.8

$$\Lambda_m \approx 10B \qquad (3.35)$$

One cannot say that the free surface width B at the bends is "always larger" or "always smaller" than that of the straight regions or of the "sections of deflections ("cross-overs" in Fig. 3.5). On the contrary, the free surface width usually tends to remain remarkably constant throughout the meandering pattern of a natural river

3.3 Curvilinear River Regions

(Fig. 3.6). Hence in the absence of detailed information, it would be only reasonable to assume that the free surface width B is the same for all sections of a meandering channel and that it can be identified with that of the straight regime channel prior to its deformation into a meandering one. Since $\overline{B} \sim Q^{\frac{1}{2}}$ (see Regime equation (3.3) and equation (3.7)) one concludes that equation (3.35) as

$$\Lambda_m \approx 10\overline{B} \sim Q^{\frac{1}{2}} \tag{3.36}$$

As is clear from Fig. 3.9a the direction of stream lines at the free surface is different from the direction of stream lines at the bed, and this is precisely the reason for the cross-sectional circulation of flow at the bends (Fig. 3.9b). It follows that the fluid (as a whole) is moving through the series of bends in a helicoidally manner, the

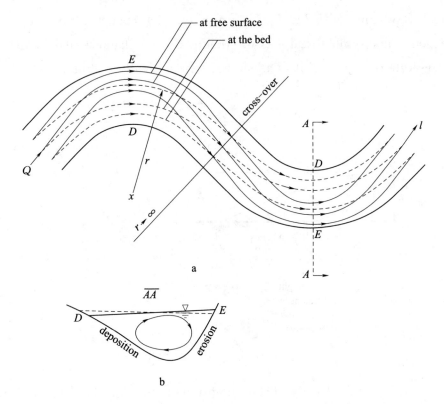

Fig. 3.9

Chapter 3 MODIFICATION OF RIVER COURSE

direction of the cross-sectional circulation alternating at each "cross-over". The free surface at the bends is not exactly horizontal: owing to centrifugal forces it is slightly inclined (higher elevation at the outer bank).

3.3.2 Cross Section Geometry of a Meandering Channel

Consider Fig. 3.9a. Since the depositions at D are mainly due to the erosions at E the cross-section area A of a meandering channel cannot be expected to vary significantly along the flow direction l (in that river region) and the experiment confirms this expectation (Rosovskii et al., 1957; Schaffernak et al., 1950; Falcon et al., 1982; etc). In contrast to this, the shape of the cross-section tends to vary significantly with the radius of curvature r of the bend (and thus with l). If r is finite ($\ll \infty$) then the flow cross-section is not symmetrical (as in straight regime channels and "cross-overs" where $r \to \infty$): it is as shown schematically in Fig. 3.10.

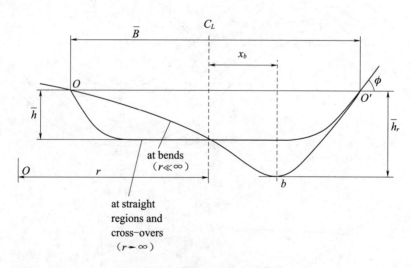

Fig. 3.10

(This has already been mentioned in 2.2.4; Fig. 2.12). Let $(\overline{h_r})$ be the (largest) bankfull flow depth at bend and $\overline{h}$ the

3.3 Curvilinear River Regions

bankfull flow depth of the corresponding straight regime channel (prior to its sinusoidal deformation). The variation of the ratio $\varepsilon = \overline{h_r}/\overline{h}$ with the relative curvature radius $r/\overline{B}$ is given by the curve in Fig. 3.11.

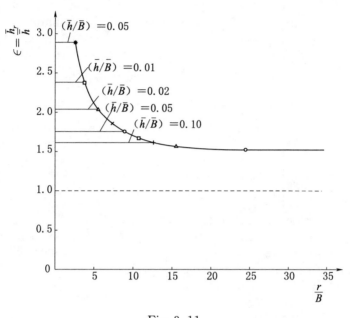

Fig. 3.11

The curve in Fig. 3.12 shows how the relative distance $x_b/\overline{B}$ (between the deepest point b and the flow center-line) varies as a function of $r/\overline{B}$. [These curves were determined from equations (16-25) of (Chow et al., 1959). This equation is due to Ripley et al. (1927)].

No agreement has been reached yet concerning the mathematical expression of the wetted perimeter $\widehat{obo'}$ of an alluvial channel at the bend. The methods proposed yield for $\widehat{obo'}$ the shapes that diverge from each other substantially (compare e.g. Fig. 16.7 in (Chow et al., 1959) with Fig. 2/4.14 in (Jansen et al., 1979)). Accordingly, for the present, perhaps the most sensible way of predicting the

Chapter 3 MODIFICATION OF RIVER COURSE

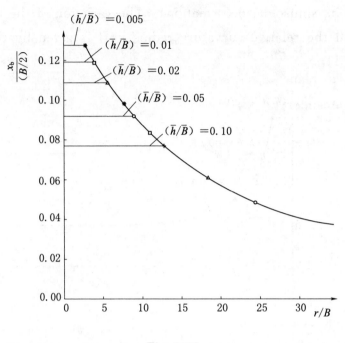

Fig. 3. 12

curve $\widehat{obo'}$, would be by adopting the following approximate method (which rests on the foregoing explanations).

(1) Using $Q=Q_{bf}$ in equation (3.3) or equation (3.7) determine $\overline{B}$ and $\overline{h}$, and thus $\overline{A}=\overline{B}\,\overline{h}$

(2) Knowing B, h, and r determine from Figs. 3.11 and 3.12 the quantities e and x_b/B and thus (h_r) and x_b.

(3) Locate the points a, b and c on a (mill metric) graph paper; take $\overline{oo'}=\overline{B}$ as horizontal. Draw the portion $\widehat{cb}$, of the curve $\widehat{abc}$, so that at c it has the angle of inclination ϕ (angle of repose) while at b it has a horizontal tangent. Draw the remaining portion $\widehat{ab}$ so that it too has at b horizontal tangent and that it has such a shape that the area of the cross-section is equal to $\overline{A}=\overline{B}\,\overline{h}$ (Fig. 3.10).

3.3.3 Modified River Channel

From experience in River Engineering it is known that if the new or modified alluvial channel (e.g. Ⅱ in Fig. 3.1) is completely straight (in plane) then usually it is not stable: if $S<S_*$ then it tends to meander; if $S=\overline{S}<S_*$ then it tends to deform as shown schematically in Fig. 3.13. The latter deformations can be regarded as "substitutes" for the prevented braiding. Considering this one tends to avoid nowadays the design of straight alluvial channels. Rather, one tends to design them so that they have a somewhat meandering form in plane. A sinusoidal form in plane tends to remain stable for any $S(>, =, <S_*)$. It seems that no rule has been proposed yet in the western world as to how this "sinusoidal form in plane" should be selected, and therefore we will adopt here the U.S.S.R. - practice (Grishin et al., 1955) which suggests that the "new" channel should have

$$\frac{a}{\Lambda_m} \approx \frac{1}{4} \quad \text{and} \quad \frac{r}{B}=4 \text{ to } 7 \qquad (3.37)$$

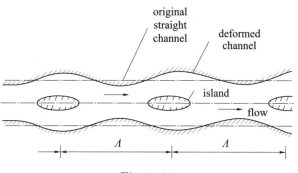

Fig. 3.13

When building the (new or modified) river channel one does not excavate it at the bends in accordance with the cross-sections shown in Fig. 3.10. One builds it simply as to have the same trapezoidal regime cross-section (corresponding to straight regions)

Chapter 3 MODIFICATION OF RIVER COURSE

throughout, and it is the flow that will eventually convert this section into that shown in Fig. 3. 10. [In other words, Fig. 3. 10 is not in order to indicate "how to build" the modified channel but to indicate "what will eventually happen" to it. In addition to this the prediction of $\overline{h_r}$ is needed also for the design of the loose revetment (see the text just below the equation (2. 39)].

Chapter 4 SPECIAL PROBLEMS

4.1 Flood Protection (Garde et al., 1977)

The flood plains of rivers are usually thickly populated and their frequent flooding results in enormous losses. Although these losses can never be evaluated accurately, the flood damage in the U.S.A. appears to range from 200 to 500 million dollars per year. Flood losses can be divided into direct and indirect losses. Direct losses include urban losses (i.e. damage to residences, industry, stores, livestock and human lives), losses in connection with transport facilities such as roads, railway lines, airports, and bridges, damages to public utilities such as schools, hospitals, water supply, and sanitation facilities, agricultural losses, etc. Indirect losses include loss of business, dislocation of public offices, stoppage of work, etc. <u>The most important aspect of flood control is the assessment of the maximum flood</u> (against which protection is to be given). In the past, structures were designed for the maximum flood on record. However, with the introduction of statistical methods for predicting the magnitude of flood discharge of various frequencies, usually a flood occurring once in a hundred years (Q_{100}) is taken as the design flood. For major structures such as spillways, 1000 years flood is normally used as the design flood. Considerable care should be exercised in extrapolation from meager data. Engineering works used for flood control can be listed under the following four categories:

(1) Flood control reservoirs (to regulate $Q=\phi_Q(t)$).

(2) Longitudinal embankments – often called levees or dikes – (to confine flow).

(3) Diversion of flow (into another channel).

(4) Channel improvement.

4.1.1 Reservoirs

A flood control reservoir (dam) is built to store water during a flood and to release it after the flood recedes. Moderation of floods depends upon the storage capacity of the reservoir. If this capacity is "small" (in relation to the drainage area) the reduction in flood discharge downstream of the reservoir may not be significant.

The cost of a reservoir built purely for flood control is usually very high and therefore in the majority of cases flood control is only one of the functions that the reservoir performs, other functions being power generation, navigation, irrigation, water supply, etc. The design of reservoirs (dams) is the topic of "Hydraulic Structures" and it will not be elaborated here.

4.1.2 Levees

Provision of levees on one or both sides of the stream to contain the flood is the oldest and most common method of flood control in use. A levee is an embankment running parallel to the stream (or nearly so) and is constructed to protect the area on one side of it from flooding. Levees are also termed as "embankments", "bunds" or "dikes". Such embankments were constructed in China as early as in 600 B.C.; levees along the Nile river in Egypt were built even earlier. In recent times, levees have been constructed on many im-

4.1 Flood Protection (Garde et al., 1977)

portant rivers of the world (Ganges and Mahanadi rivers in India, Changjiang, Huaihe, and Huanghe rivers in China, Mississippi river in U. S. A., Po, Danube, and Rhone rivers in Europe). From a historical viewpoint all the great levee systems in the world have the following aspects in common:

(1) Levees have been extended gradually.

(2) There has been a gradual enlargement in the cross-section of levees.

(3) None of the levee systems has been free from breaches.

Levees can be extended gradually (in the flow direction l) to protect the progressively increasing area from floods. Also, since locally available materials and labor are utilized (in the construction of levees) they are fairly inexpensive and simple. As against these advantages, there are also a few disadvantages. Levees being made up of earth are susceptible to boring action by animals and thus vulnerable to "piping failure". Levee breaches, especially in the upper reaches, can result in flooding of the entire area which depends on levees for protection. For these reasons, levees need very careful supervision especially during floods and any breaches need to be plugged almost on a war-footing. Generally, levees of height greater than about 15 m are uneconomical.

(1) Effect of levees on rive regime: In the absence of levees, the flood water spills over the natural river banks and spreads over the flood plain. Thus the flow width increases, the flow velocity decreases, and consequently most of the suspended load is deposited on the flood plain. Very often this makes the soil of the flood plain more fertile and improves agriculture (Nile river - throughout history). When levees are constructed, the width of the stream is reduced, the velocity increased, and the rate of deposition on the flood plain becomes smaller. The material that should have deposited

on the flood plain in the absence of levees is now carried downstream (and deposited either in the unlevied region of the flood plain or at the River-end in the sea).

Other effects of compelling the flood water to flow between the embankments are:

(a) Increase in the speed of the propagation of flood wave.

(b) Rise in the free surface elevation of the river during the flood.

(c) Reduction of the "storage" and thus an increase of the maximum flow rate.

(d) Decrease in the water surface slope of the river immediately upstream of the leveed region (as a result aggradation may take place in that region).

(2) Alignment and design of levees: The alignment of levees depends on the location of important industries, towns and other areas along the stream that need to be protected. If levees on the two banks are located sufficiently close to each other, the entire flood plain could be protected from floods; but in such cases, levees will be very high and massive and therefore very expensive. On the other hand, the levees on the left and right flood plains cannot be arbitrarily close to each other for the distance between them cannot be less than the width of the "meander belt". The levees should also have the general curvatures (in plane) that would be approximately the same as those of the river flowing (during the flood) on flood plains.

Even a small increase in the height of the levee makes its cost increase appreciably, and therefore the height of a levee must be determined (for the adopted Q_M with the aid of model tests. The level of the top of a levee can be fixed after providing a freeboard 1m to 1.5m. (The freeboard is a provision against waves and the

4.1 Flood Protection (Garde et al., 1977)

possibility of occurrence of a flood of higher return period than assumed.) A probable settlement of the levee after its construction must also be considered.

The top width of levees should be adequate for, at least, a small vehicle to move on it; the top width should not be less than 3m. The design of a levee follows the same principles as those for the design of an earth dam. However, while the stream face of an earth dam is exposed to water (static) most of the time, that of a levee is exposed to (flowing) water only for a short period during a year. The river-side slope of levees varies from 1/2 to 1/6 (1/3 is probably the most favored slope). If the levee is higher than about 5m or so, then it is common practice to provide it with berms (of adequate width) at intermediate elevations. Fig. 4.1 shows cross-sections of some of the existing levees.

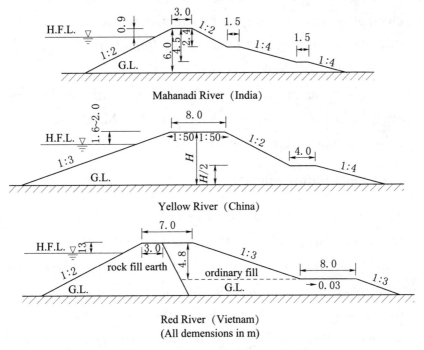

Fig. 4.1 Typical cross sections of levees

Chapter 4 SPECIAL PROBLEMS

4.1.3 Flow Diversion

One of the most effective and economic ways of flood control is the diversion of a part of the flood flow rate from the main channel. The diverted water can pass through either a natural or an artificial channel and then enter the sea or a lake. In some cases, the diverted water joins the main channel again some distance downstream.

Some permanent diversion devices have been constructed in the lower Mississippi River (U.S.A.). The design flood flow rate of the Mississippi River at Natchez is $Q_M = 3,030,000$ cfs. Out of this 620,000 cfs is diverted through the old river and 600,000 cfs through the Morganza floodway (Fig. 4.2). Both of these diverted flows merge into the Atchafalaya River which discharges into the Gulf of Mexico. Of the remaining 1,500,000 cfs that flows in the main channel of the Mississippi (downstream the Morganza floodway) 250,000 cfs is diverted to the Lake Pontchartrain and thereby also to the Gulf of Mexico (through Bonnet Carre spillway): 1,250,000 cfs is left in the river to flow to the Gulf. Similarly water is diverted from the Assiniboine River to Lake Manitoba to protect the city of Winnipeg (Kuiper et al., 1965).

4.1.4 Channel Improvement

(1) A river channel can be improved by <u>reducing its roughness</u> and consequently by conveying the flood flow rate with smaller flow depths and thus with reduced elevations of the free surface. The reduction of roughness can be achieved by removing weeds, trees, and other obstacles from the banks and the flood plain.

(2) River channels can also be improved by <u>increasing their cross - section</u> (by widening and deepening them with the aid of

4.1 Flood Protection (Garde et al., 1977)

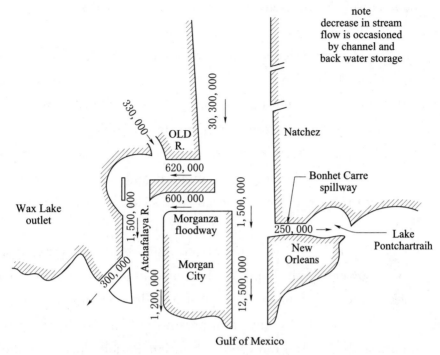

Fig. 4.2 Flow diversion on the Mississippi river

dredging). The material removed by dredging may be used for filling secondary channels or for building or strengthening levees, dikes, etc. Usually, it is not economic to dredge the river channel over a long reach.

(3) One of the most important methods of channel improvement is channel shortening by executing cutoffs (see 2.3.3). If the length between two sections along a channel is reduced by cutting off bends, the slope and consequently the flow velocity is increased while flow depths at the elevations of the free surface are decreased. Thus sixteen cutoffs were executed on the Mississippi River (between 1933 and 1942) thereby the original river length of 452 miles (between Memphis and Baton Rouge) was reduced by 152 miles. As a result, the flood water levels were lowered by 7 to 8 ft at Vicksburg and 12 to 13 ft at Arkansas City. On a smaller scale

· 77 ·

Chapter 4 SPECIAL PROBLEMS

in Canada cutoffs were used, e. g. to reduce the floodwater levels of the Pembina River in Alberta (Blench et al. , 1966).

The human interference to the Mississippi river in general and the shortening of its length by a series of "cut - off" channels, in particular, drew the attention of the great American writer Mark Twain who made the following comment on the topic.

The Mississippi between Cairo and New Orleans was twelve hundred and fifteen miles long one hundred and seventy - six years ago; its length is only nine hundred and seventy - three miles at present.

Please observe: in the space of one hundred and seventy - six years, the Lower Mississippi has shortened itself two hundred and forty - two miles. That is an average of a trifle over one mile and a third per year. Therefore, any calm person, who is not blind or idiotic, can see that in the old Oolitic Silurian Period, just a million years ago next November, the Lower Mississippi River was upward of one million three hundred thousand miles long, and stuck out over the Gulf of Mexico like a fishing - rod. And by the same token, any person can see that seven hundred and forty - two years from now the Lower Mississippi will be only a mile and three - quarters long, and Cairo and New Orleans will have joined their streets together, and be plodding comfortably along under a single mayor and a mutual board of aldermen. Something is fascinating about science. One gets such wholesale returns of conjecture out of such a trifling investment of fact.

4.2 Degradation and Aggradation (Yalin et al. , 1983)

If the erosion of a river bed is of local character then it is termed as "scour" (scour at a bridge pier, scour at the stilling basin, etc).

4.2 Degradation and Aggradation (Yalin et al., 1983)

If the erosion of a river bed extends over a "long" distance (several kilometers) then it is called the "degradation" (of the river bed). "Aggradation" is the name of the opposite process, i.e. of the deposition that extends over a "long" distance. Consider, for example, the influence of a (recently erected) dam on the river bed (Fig. 4.3).

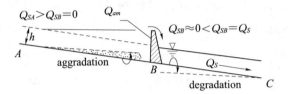

Fig. 4.3

Flow rate Q "passes" through the dam, yet sediment is "trapped" by it. Hence at the dam-section B, we have $Q_S = 0$ (or almost zero) whereas in the unaffected river regions (downstream of C and upstream of A) Q_S has a certain finite value. It follows (from Exner Polya equation (1.4)), that

(1) along $\overline{BC}$, where

$$\frac{\partial q_s}{\partial l} \approx \frac{(Q_S - B) - 0}{L} > 0 \qquad (4.1)$$

we have $\partial z / \partial t < 0$, i.e. we have the degradation; whereas

(2) along $\overline{AB}$, where

$$\frac{\partial q_s}{\partial l} \approx \frac{0 - (Q_S/B)}{L} < 0 \qquad (4.2)$$

we have $(\partial z/\partial t) > 0$, i.e. we have the aggradation Similarly, consider a clear water channel that is (recently) connected to the river (Fig. 4.4a). The combined flow rate $Q + Q'$ is "stronger" than the usual river flow rate Q and thus it is likely to induce a larger Q_S than that induced by Q. But if so, then a river region ($\overline{BC}$, say) may be subjected to degradation. In contrast to this, the construction of a clear water intake from the river (Fig. 4.4b), is likely to induce aggradation along with a river region AB (for Q_S at A is

Chapter 4 SPECIAL PROBLEMS

larger than that at B where the transporting flow rate is reduced. Many other examples that may induce the degradation or aggradation processes can be shown. Yet, for the sake of clarity, in the following, these processes will be analyzed for the case of a dam (Fig. 4.3) only. It will be assumed that the dam acts as a perfect "sediment trap" ($Q_S=0$ at the dam-section) and that the river aspect ratio B/h is larges (and thus that the flow can be treated as two dimensional). Furthermore, it will be postulated that both degradation and aggradation extend over a certain known distance L. [e.g. in the case of degradation, the section C can be a non-erodible rocky sill (which extends over the whole river width), it can be the river mouth or it can be the location of the upstream end of the storage (reservoir) created by yet another dam downstream of C. In the case of aggradation, the section A can be identified with the approximate beginning of the backwater curve (retarded flow) caused by the dam at B].

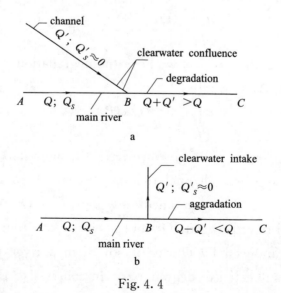

Fig. 4.4

4.3 Mathematical Formulation

(1) Differential Equation

The transport rate q_s can be given as the product of the weight G of the grains transported over the unit area of the bed with the migration velocity V of that weight:

$$q_s = GV \tag{4.3}$$

The weight G can be expressed as

$$G \sim N\gamma_s D^3 \tag{4.4}$$

Where $\gamma_s D^3$ reflects the submerged weight of a single grain and N is the number of grains transported over the unit area of the bed. The number N can depend only on how the "fluid action" τ_0 on the bed grains compares with their "resistance" (their weight per unit area of the bed) $\gamma_s D$; i.e. on the mobility number $Y = \tau_0/\gamma_s D$. No grains are lifted when $Y < Y_{cr}$, and therefore N must be an increasing function of $Y - Y_{cr}$ It follows that $N \sim f(Y - Y_{cr})/D^2$ and thus

$$G \sim \gamma_s D f(Y - Y_{cr}) \tag{4.5}$$

Since the migration of grains is induced by the flow, it would be only reasonable to consider V as proportional to the flow velocity v:

$$V \sim v \tag{4.6}$$

Substituting equation (4.5) and equation (4.6) in equation (4.3) one determines for q_s

$$q_s \sim \gamma_s D v f(Y - Y_{cr}) \tag{4.7}$$

It seems that among the prominent relations it is only Bagnold's bed-load formula

$$\phi = \rho^{\frac{1}{2}} q_s / (\gamma_s D)^{\frac{3}{2}} = 4.25 Y^{\frac{1}{2}}(Y - Y_{cr}) \tag{4.8}$$

which can be expressed as

$$q_s = 4.25 v_* [\tau_0 - (\tau_0)_{cr}] \tag{4.9}$$

Chapter 4 SPECIAL PROBLEMS

and thus

$$q_s = \frac{4.25}{c} \gamma_s D v (Y - Y_{cr}) \qquad (4.10)$$

which is consistent with equation (4.7). Here c is the friction factor and $f(Y-Y_{cr})$ is a linear function:

$$f(Y - Y_{cr}) \sim (Y - Y_{cr}) \qquad (4.11)$$

Introducing the volumetric transport rate q_s/γ_s, bearing in mind that $vh = q$, and taking into account that

$$Y = \frac{\gamma_s h}{\gamma_s D} \quad \text{and} \quad Y_{cr} = \frac{\gamma_{s_{cr}} h_f}{\gamma_s D} \qquad (4.12)$$

where h and h_f are flow depths when the river bed slope is S and S_{cr} respectively, one can express equation (4.10) as follows

$$\frac{q_s}{\gamma_s} = \beta \left[S - \frac{h_f}{h} S_{cr} \right] \qquad (4.13)$$

Here

$$\beta = \frac{4.25}{c} \frac{\gamma}{\gamma_s} q \qquad (4.14)$$

Let x be the coordinate that coincides with the direction of river flow and which reflects the location (distance) of a river section with respect to the darn (responsible for degradation and/or aggradation processes). Clearly, in the course of these processes, the elevation z of the river bed varies as a function of the location x and the time $t [z = f(x, t)]$, the slope of the river bed is

$$S = -\frac{\partial z}{\partial x} \qquad (4.15)$$

(minus sign because z decreases when x increases (in the direction of Flow)).

We will assume that the flow conditions are (nearly) uniform in the river region (L) where degradation or aggradation takes. place (which means that q, c, B, h may vary with t but not with x). But if so, then

$$\frac{\partial(q_s/\gamma_s)}{\partial_x} - \frac{\partial}{\partial_x}\left[\beta - \left(\frac{\partial z}{\partial x} - \frac{h_f}{h}S_{cr}\right)\right] = -\beta\frac{\partial^2 z}{\partial x^2} \qquad (4.16)$$

Substituting this value in the Exner Polya equation

$$\frac{\partial q_s}{\partial x} + \gamma_s \frac{\partial z}{\partial t} = 0 \qquad (4.17)$$

i. e.

$$\frac{\partial(q_s/\gamma_s)}{\partial x} = -\frac{\partial z}{\partial t}$$

We arrive at the differential equation

$$\frac{\partial^2 z}{\partial x^2} = \frac{1}{\beta}\frac{\partial z}{\partial t} \qquad (4.18)$$

which is the same as that of heat transfer and diffusion.

Now we go over to the presentation of the solutions of this equation (which correspond to a two-dimensional flow and to $B =$ const (i. e. to $c =$ const and $q =$ const)).

(2) Degradation

We adopt the notation shown in Fig. 4.5.

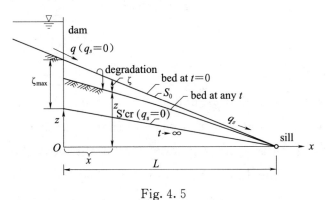

Fig. 4.5

The functioning of the dam, and thus the supply of the clear fluid ($q_s = 0$) at the section $x = 0$ initiates at $t = 0$. When $t = 0$ then the river bed slope in the region $0 < x < L$ is $S = S_0$. The transport of bed material (in the degradation region $0 < x < L$) and consequently the degradation process terminates when the bed slope is reduced to

Chapter 4 SPECIAL PROBLEMS

$S = S_0$ (all over the region $0 < x < L$). i.e. the initial and boundary conditions are

$$t = 0 : S = S_0 \text{ and thus } z = S_0(L-x) \atop t \to \infty : S = S_{cr} \text{ and thus } z = S_{cr}(L-x)\} \quad (4.19)$$

the solution of equation (4.19) corresponding to these conditions (and B = const)

$$\frac{\zeta}{\zeta_{max}} = (1-\xi) - \frac{8}{\pi^2} \sum_{n=0}^{\infty} \left\{ \frac{\cos\left[(2n+1)\frac{\pi}{2}\xi\right]}{(2n+1)^2} \exp\left[-\frac{(2n+1)^2 \pi^2}{4}\right] \right\} \quad (4.20)$$

here

$$\theta = \beta \frac{t}{L^2} = 4.25 \frac{q \gamma}{c \gamma_s} \frac{t}{L^2} \quad \xi = \frac{x}{L} \quad (4.21)$$

and

$$\xi_{max} = (S_0 - S_{cr})L \quad (4.22)$$

Furthermore, for the final (degraded) bed we have

$$q = h_f c \sqrt{g S_{cr} h_f} \quad (4.23)$$

and

$$\frac{\gamma}{\gamma_s} \frac{S_{cr} h_f}{D} = Y_{cr} \quad (4.24)$$

(it is assumed that q and c (and thus β) remain constant throughout the degradation process and that h varies only because S varies). As is clear from equation (4.20 the degradation process terminates (i.e. the left – hand side of equation (4.20) becomes unity for $x=0$) only when $\theta \sim t \to \infty$ (as indicated in equation (4.19)). Hence (in analogy to "terminal velocity", "boundary layer", "normal probability distribution" etc) it must be assumed here that from the practical standpoint the degradation process terminates when t' has reached such a value T which corresponds to, say, $\zeta = 0.95 \zeta_{max}$ at $x=0$. i.e. if equation (4.20) is symbolized by

4.3 Mathematical Formulation

$$\frac{\zeta}{\zeta_{max}} = f(x,t) \qquad (4.25)$$

then the "practical duration of degradation" T is given by

$$0.95 = f(0,T) \qquad (4.26)$$

(3) Aggradation

We adopt the notation shown in Fig. 4.6. This notation is essentially the same as in Fig. 4.5. The difference is that stands for the "amount of aggradation" (rather than for the "amount of degradation" as in Fig. 4.5) corresponding to a location x at the time t.

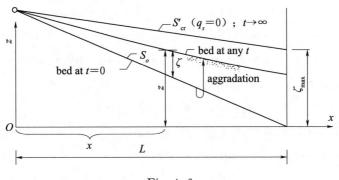

Fig. 4.6

Furthermore, the origin of the x-axis is altered ($x = 0$ at the "zero aggradation section", $x = L$ at the dam). This alteration of the origin permits the identification of the present initial and boundary conditions with equation (4.19), the solution of equation (4.18) corresponding to the case of aggradation (taking place for β = const) being thus

$$\frac{\zeta}{\zeta_{max}} = \xi - \frac{8}{\pi^2}\sum_{n=0}^{\infty}\left\{\frac{\cos\left[(2n+1)\frac{\pi}{2}(1-\xi)\right]}{(2n+1)^2}\exp\left[-\frac{(2n+1)^2\pi^2}{4}\theta\right]\right\} \qquad (4.27)$$

Here ζ, θ and ζ_{max} are given by exactly the same relations as equation (4.21) and equation (4.22). Equation (4.24) are also valid as they stand; the "practical duration of aggradation" is also

Chapter 4 SPECIAL PROBLEMS

given by equation (4.26) (assuming, of course, that equation (4.25) symbolizes equation (4.27)).

(4) Generalization of the Equation (4.18)

In practice, the order of magnitude of T ("practical duration of aggradation or degradation processes") is several years. Thus the solutions presented above may deviate from the truth substantially if e.g. Q varies substantially with the time of the year; i.e. if $Q = \Psi(t)$, and thus β, can hardly be regarded as constants. In most of thepractical cases

$$Q = \Psi(t) \quad B = \Psi(x) \quad c = f(x) \tag{4.28}$$

and it would be worthwhile to investigate how the differential equation (4.18) can be generalized as to cover such cases. We assume that B/h is sufficiently large, and thus that

$$Q = qB \quad \text{and} \quad Q_s = q_s B \tag{4.29}$$

are valid for all locations x of the region $0 < x < L$. In this case, the continuity of sediment transport implies

$$\frac{\partial Q_s}{\partial x} = -\gamma_s \frac{\partial (zB)}{\partial t} \tag{4.30}$$

i.e.

$$\frac{\partial (Q_s/\gamma_s)}{\partial x} = -B \frac{\partial z}{\partial t} \tag{4.31}$$

which is the generalized Exner Polya equation. Now

$$(Q_s/\gamma_s) = (q_s/\gamma_s)B = -\frac{4.25}{c}\frac{\gamma}{\gamma_s}q\left[\frac{\partial z}{\partial x}+S_{cr}\right]B = -4.25\frac{\gamma Q}{\gamma_s c}\left[\frac{\partial z}{\partial x}+S_{cr}\right] \tag{4.32}$$

which gives

$$\frac{\partial (Q_s/\gamma_s)}{\partial x} = -4.25\frac{\gamma}{\gamma_s}Q\frac{\partial}{\partial x}\left[\frac{1}{c}\left[\frac{\partial z}{\partial x}+S_{cr}\right]\right] \tag{4.33}$$

Substituting equation (4.33) in equation (4.31) we obtain

$$-\frac{1}{c^2}\frac{dc}{dx}\left[\frac{\partial z}{\partial x}+S_{cr}\right]\frac{1}{c}\frac{\partial^2 z}{\partial x^2} = \left[4.25\frac{\gamma Q}{\gamma_s B}\right]^{-1}\frac{\partial z}{\partial x} \tag{4.34}$$

i. e.

$$-\frac{1}{c}\frac{dc}{dx}\left[\frac{\partial z}{\partial x}+S_{cr}\right]+\frac{\partial^2 z}{\partial x^2}=\frac{1}{\beta_*}\frac{\partial z}{\partial t} \quad (4.35)$$

where

$$\beta_* = \frac{4.25\,\gamma}{c}\frac{Q}{\gamma_s B} = \frac{4.25\,\gamma}{c\,\gamma_s}q \quad (4.36)$$

but this time it is a variable quantity (for Q, B and c are given equation (4.28) [It is interesting to point out that if the variation of the friction factor c with the location x is neglected (i. e. if $(dc/dx)=0$ then equation (4.35) reduces to equation (4.18); with the difference that β is a constant whereas β_* is a given function of x and t.]

Knowing the functions Q, B, and c, the variation of z with x and t can be computed from equation (4.35) by numerical methods. Another approach, and at the present perhaps the most reliable one, is by using a physical model (next paragraph).

(5) Dynamic Similarity

Consider equation (4.35) and equation (4.36). Let Q_0, B_0 and c_0 some typical (constant) values of Q, B, and c. In this case β_* can be expressed as follows

$$\beta_* = \beta_0 \frac{Q}{Q_0}\frac{B_0 C_0}{B\,C} \quad (4.37)$$

where

$$\beta_0 = \frac{4.25\,\gamma}{c_0}\frac{Q_0}{\gamma_s B_0} \quad (4.38)$$

Introducing the dimensionless variables

$$\xi = \frac{x}{L} \quad \text{and} \quad \theta = \frac{\beta_0}{L^2}t \quad (4.39)$$

one can express the given functions Q, B and C in the dimensionless form:

$$\frac{Q}{Q_0}=\phi_*(\theta) \quad \frac{B_0}{B}=\Psi_*(\xi) \quad \frac{C_0}{C}=f_*(\xi) \quad (4.40)$$

Chapter 4 SPECIAL PROBLEMS

Let z_0 be a typical elevation and η the dimensionless counterpart of z:

$$\eta = \frac{z}{z_0} \tag{4.41}$$

The relations equation (4.37) to equation (4.41) make it possible to express the differential equation (4.35) in the following dimensionless form

$$\frac{1}{f_*(\xi)} \frac{df_*(\xi)}{d\xi} \left[\frac{\partial \eta}{\partial \xi} + \frac{L}{z_0} S_{cr} \right] + \frac{\partial^2 \eta}{\partial \xi^2} = \frac{\Psi_*(\xi) f_*(\xi)}{\phi_*(\theta)} \frac{\partial \eta}{\partial \theta} \tag{4.42}$$

which can be shown symbolically as

$$F_1(\xi) \left[\frac{\partial \eta}{\partial \xi} + A \right] + \frac{\partial^2 \eta}{\partial \xi^2} = F_2(\xi, \theta) \frac{\partial \eta}{\partial \xi} \tag{4.43}$$

where $F_1(\xi)$ and $F_2(\xi, \theta)$ are given (known) functions and $A = S_{cr} L / z_0$ is a constant. Clearly for a given geometry of the boundary conditions the solution of this dimensionless differential equation (i.e. the function $\eta = F(\xi, \theta)$) depends entirely on the numerical value of the constant A; and if A remains unchanged then the function $\eta = F(\xi, \theta)$ this equation will yield will also remain unchanged. But this means that the dynamic similarity is determined by

$$\left. \begin{array}{ll} \lambda_\xi = 1 & \text{i.e.} \quad \lambda_x = \lambda_L \\ \text{and} \quad \lambda_\theta = 1 & \text{i.e.} \quad \lambda_t = \lambda_L^2 / \lambda_{B_0} \\ \lambda_A = 1 & \text{i.e.} \quad \lambda_{S_{cr}} = \lambda_{z_0} = \lambda_L \end{array} \right\} \tag{4.44}$$

The first condition is trivial as it merely indicates that all horizontal lengths must be of the same scale. The second condition supplies the time scale of the degradation (or aggradation) process:

$$\lambda_t = \lambda_L^2 \lambda_{B_0}^{-1} = \lambda_x^2 \lambda_c \lambda_{\gamma S} \lambda_B \lambda_\gamma^{-1} \lambda_Q^{-1} \tag{4.45}$$

and since

$$\lambda_Q = \lambda_v \lambda_h \lambda_B \quad \text{and} \quad \lambda_c = \lambda_v \lambda_{v_*}^{-1} \tag{4.46}$$

we obtain

$$\lambda_t = \lambda_x^2 \lambda_{\gamma S} \lambda_h^{-1} \lambda_\gamma^{-1} \lambda_{v_*}^{-1} \tag{4.47}$$

Now, for a practical model, we have $\lambda_\gamma = 1$; while λ_{v_*} is given by

$$\lambda_{v_*} = \lambda_h^{\frac{1}{2}} \lambda_S^{\frac{1}{2}} = \lambda_y \lambda_x^{-\frac{1}{2}} \qquad (4.48)$$

hence

$$\lambda_t = \frac{\lambda_x^{2.5}}{\lambda_y^2} \lambda_{ys} \qquad (4.49)$$

Consider finally the third condition in equation (4.44). Since $\lambda_{z_0} = \lambda_y$ (vertical model scale) while $\lambda_L = \lambda_x$ we have $\lambda_{S_{cr}} = \lambda_y / \lambda_x$. The value of S_{cr} is given by

$$Y_{cr} = \frac{\gamma h S_{cr}}{\gamma_S D} = 0.05 \qquad (4.50)$$

which must be identical in model and prototype. From this identity, i.e. from

$$\lambda_{Y_{cr}} = 1 \quad \text{we obtain} \quad \lambda_{S_{cr}} = \frac{\lambda_{Y_S} \lambda_D}{\lambda_\gamma \lambda_h} = \frac{\lambda_{\gamma_S} \lambda_D}{\lambda_y} \qquad (4.51)$$

and equating this value with λ_y / λ_x we arrive at

$$\lambda_{\gamma_S} \lambda_D = \frac{\lambda_y^2}{\lambda_x} \qquad (4.52)$$

The equation (4.49) and equation (4.52) form the required system of scale relations needed for the design of a distorted model for the study of degradation or aggradation processes. These two equations involve five scales (λ_x, λ_y, λ_t & λ_{γ_S}, λ_D), and therefor only three of them can be selected freely.

(6) Discussion

Observe, that the fulfillment of the condition equation (4.51) implies automatically that the values of the Y number are identical in model and prototype for all stages and locations; i.e. that

$$\lambda_Y = 1 \quad \text{and thus} \quad \lambda_S = \frac{\lambda_{\gamma_S} \lambda_D}{\lambda_h} \quad (=, \lambda_y / \lambda_x) \qquad (4.53)$$

are valid for all x and t. But the model and prototype identity of Y and Y_{cr} implies, as can be inferred from equation (4.8), the model and prototype identity of ϕ; i.e. it signifies the (necessary)

Chapter 4 SPECIAL PROBLEMS

similarity of the bedload motion [as given by Bagnold's formula having the form $\phi = f(Y, Y_{cr})$]. To achieve the similarity of sediment transport in general (i. e. as to include suspended load, viscous influences, etc) one needs to have the identity of $\phi = f(Y, Y_{cr}, X, Z, \cdots)$, i. e. one needs to satisfy [in addition to equation (4.49) and equation (4.52)] also the conditions

$$\lambda_X = 1, \quad \lambda_Z = 1, \cdots \tag{4.54}$$

But this may turn out to be very difficult, or even impossible, for the case of a conventional physical model operating with the prototype fluid (i. e. for the case $\lambda_Y = 1$ and $\lambda_v = 1$).

A very elegant mathematical approach to the study of degradation has been developed also by K. C. Wilson. In this approach the degradation phenomenon is formulated without the need to invoke the "distance to the sill" L.

REFERENCES

[1] JANSEN P H, VAN BENDEGOM, VAN dEN BERG J, et al., 1979. Principles of River Engineering [J]. Pitman, London.

[2] LEOPOLD L B, et al., 1964. Fluvial Processes in Geomorphology [J]. W. H. Freeman and Co, San Francisco &. London.

[3] GARDE R J, RANGA RAJU, K. G., 1977. Mechanics of Sediment Transportation and Alluvial Stream Problems [J]. Wiley Eastern Ltd.

[4] WEN SHEN H, 1971. River Mechanics [J]. Vol and II, Colorado State University, Fort Collins, Colorado.

[5] NEILL C R, 1977. Short Course on River Mechanics [J]. Fac. of Extension, Univ. of Alberta, Edmonton, Alta.

[6] HENDERSON F M, 1966. Open Channel Flows [J]. MacMillan, New York.

[7] MORRIS H M, 1963. Applied Hydraulics in Engineering [J]. Ronal Press Co, New York.

[8] SIMONS D B, SENTURK, F., 1977. Sediment Transport Technology [J]. Water Resources Publications, Fort Collins, Colorado.

[9] EINSTEIN H A, 1964. River Sedimentation [M]. Handbook of Appl. Hydrology (Ven Te Chow, Ed.), McGraw-Hill, Section 17-11.

[10] CHOW, VEN TE, 1959. Open Channel Hydraulics [J]. McGraw-Hill, New York.

[11] WEN SHEN H, 1971. Sedimentation [J]. (H. A. Einstein) Colorado State University, Fort Collins, Colorado.

[12] WHITE C M, 1940. The Equilibrium of Grains on the Bed of a Stream. [J]. Proceedings of the Royal Society of London. Series A, Mathematical and Physical Sciences (1934—1990), 174 (958).

REFERENCES

[13] ROSOVSKII I L, 1957. Flow of Water in Bends of Open Channels [J]. Academy of Sc. of the Ukrainian S. S. R. , Kiev.

[14] STEPHENSON D, 1977. Rockfill in Hydraulic Engineering [J]. Elsevir, Amsterdam, Oxford.

[15] ODGAARD A J and KENNEND J F, 1982. Analysis of Sacramento River Bend Flows and Development of a New Method for Bank Protectionz [J]. I. I. H. R. Rep. N0 241, Iowa Institute of Hydraulic Research, Iowa City.

[16] VAN DER LEEDEN F, 1975. Water Resources of the World [J]. Water Information Center, . Port Washington, New York.

[17] BULLOCK A, 1965. Great Rivers of Europe [J]. Weidenfeld and Nicholson, London.

[18] Rand McNally Encyclopedia, 1980. World Rivers [M]. Rand McNally & Co, Chicago - New York.

[19] MacLennan H, 1974. Rivers of Canada [J]. MacMillan of Canada, Toront.

[20] Canada Water Year Book [M]. Environment Canada, 1975.

[21] Research Council of Alberta, 1970. Selected Characteristics of Stream Flow in Alberta [J]. Edmonton, Alta.

[22] Research Council of Alberta, 1972. Hydraulic and Geomorphic Characteristics of Rivers in Alberta [J]. Edmonton, Alta.

[23] Edmonton, Alta, 1972. River Engineering and Hydrology for the Pipe - Iine Industry [J]. Northwest Hydraulic Consultants Ltd.

[24] M NIXON, 1959. A Study of the Bank - Full Discharges of Rivers in England and Wales [J]. Proceedings of the Institution of Civil Engineers, 12 (2).

[25] LEWIN J, 1981. British Rivers [J]. George Allen and Unwin, London.

[26] BLENCH T, 1966. Mobile Bed Fluviology [J]. T. Blench & Associates Ltd, Edmonton, Alberta.

[27] BLENCH T, 1957. Regime Behaviour of Canals and Rivers [J]. Butterworths Scientific Publications, London.

[28] SIMONS D B, AJBERTSON M L, 1960. Uniform Water Conveyance Channels in Alluvial Material [J]. Proc. A. S * C. E. , Vol. 86, HY5.

[29] FRIEDKIN J F, 1945. A Laboratory Study of the Meandering of Alluvial Rivers [J]. U. S. W. E. S. , Vicksburg, Miss.

REFERENCES

[30] INGLIS C C, 1949. The Behaviour and Control of Rivers and Canals [J]. Pt. I, CWINRS, Research Publication No. 13.

[31] BOGARDI J, 1974. Sediment Transport in Alluvial Streams [J]. Akadémiai Kiado, Budapest, Hungary.

[32] RAUDKIVI A J, 1976. Loose Boundary Hydraulics [M]. (Second Edition), Pergamon Press, Oxford, England.

[33] GRAF, W H, 1971. Hydraulics of Sediment Transport [J]. McGraw-Hill.

[34] YALIN M S, 1977. Mechanics of Sediment Transport [M]. (Second Edition), Pergamon Press, Oxford, England.

[35] YALIN M S, 1957. Die. Theoretishe Analyse der Mechanik der Geschiebebewegun [J]. Mitteilungsblatt der BAW, No. 8.

[36] EINSTEIN H A, 1950. The Bed Load Function for Sediment Transportation in Open Channel Flows [J]. U. S. Dept. Agriculture, Soil Conserv. Ser., Tech. Bull. 1026.

[37] DUHM J, 1951. Wasserbau [M], 1. Teil: Der Flussbau, Verlag G. Fromme & Co, Wien.

[38] PRESS H, 1956. Binnenwasserstrassen und Binnenhafen [J]. William Ernst & Sohn, Berlin.

[39] SCHAFFERNAK F, 1950. Gyrundriss der Flussmorphologie und des Flussbaues [J]. Springer-Verlag, Wien.

[40] GRISHIN M M, 1955. Hydrotechnical Structures [J]. (in Russian), State Publishing House for Civil Engineering and Architecture, Moscow.

[41] ZAMARIN E A, Popov K V, Fandeev V V, 1952. Hydrotechnical Structures [J]. (in Russian) State Publishing House for Agricultural literature, Moscow.

[42] MAKKAVEEV N I, 1955. River Bed and Erosion Processes [J]. Academy of Sciences of the U. S. S. R., Moscow.

[43] ZNAMENSKAYA N S, 1976. Sediment Transport and Alluvial Processes [J]. Hydrometeoizdat, Leningrad.

[44] LEVI I I, 1957. Dynamics of Alluvial Streams [J]. State Energy Publishing, Moscow, Leningrad.

REFERENCES

[45] ANDERSON A G, PAINTAL A S, DAVENPORT, J T, 1970. Tentative Design Procedures for Riprap - Lined Channels [J]. National Cooperative Highway Research Program, Report 108, Highway Research Board, National Research Council.

[46] El-KHUDAIRY M, 1970. Stable Bed Profiles in Continuous Bends [J]. Thesis presented to the University of California, at Berkeley, California, in 1970, in partial fulfillment of the requirements for the degree of Doctor of Philosophy.

[47] ENGELUND F, 1974. Flow and Bed Topography in Channel Bends [J]. Journal of the Hydraulics Division, ASCE, VOL. 100, NO. HY11, Proc. Paper 10963, pp. 1631-1648.

[48] ENGELUND F, 1974. Experiments in Curved Alluvial Channel [R]. Institute of Hydrodynamics and Hydraulic Engineering, The Technical University of Denmark, Progress Report No. 34.

[49] ENGELUND F, 1976. Experiments in Curved Alluvial Channel (Part 2) [R]. Institute of Hydraulic Engineering, The Technical University of Denmark, Progress Report No. 38.

[50] ENGELUND F, Jørgen Fredsøe, 1976. A Sediment Transport Model for Straight Alluvial Channels [J]. Hydrology Research, 7 (5).

[51] FALCON M A, 1979. Analysis of Flow in Alluvial Channel Bends [J]. Thesis presented to The University of Iowa, Iowa City, Iowa, in partial fulfillment of the requirements for the degree of Doctor of Philosophy.

[52] FALCON, M A, Kennedy J F, 1982. Flow in Alluvial - River Curves [J]. Journal of Fluid Mechanics.

[53] IDEDA S, 1974. On Secondary Flow and Dynamic Equilibrium of TransverseBed Profile in Alluvial Curved Open Channel [J]. Proceedings of the Japanese Society of Civil Engineers, No. 229, pp. 55-65.

[54] KIKKAWA H, IKEDA S, KITAGAWA A, 1976. Flow and Bed Topography in Curved Open Channels [J]. Journal of the Hydraulics Division, ASCE, Vol. 102, No. HY9, Proc. Paper 12416, pp. 1327-1342.

[55] ODAGAARD A J, 1981. Transverse Bed Slope in Alluvial Channel Bends [J]. Journal of the Hydraulics Division, Proceedings ASCE, Vol. 107,

REFERENCES

No. Hl2.

[56] ZIMMERMANN C, KENNEDY J F, 1978. Transverse Bed Slopes in Curved Alluvial Streams [J]. Journal of the Hydraulics Division, ASCE, Vol. 104, No. HY1, pp. 33 – 48.

[57] VIESSMAN W Jr, et al, 1977. Introduction to Hydrology [J]. Second Edition, Harper and Row, New York, London.

[58] IZBASH S V, KHALDRE, Kh Yu, 1970. Hydraulics of River Channel Closure [J]. Transl. Cairns, G. L. , Butterworths, London.

[59] NAYLOR A H, 1976. A Method for Calculating the Size of Stone needed for Closing End Tipped Rubble Banks in Rivers [J]. CIRIA Rep. London.

[60] RIPLEY H C, 1927. Relation of Depth to Curvature of Channels [J]. Transactions A. S. C. E. , Vol 90.

[61] KUIPER E, 1965. Water Resources Development – planning, Engineering and Economics [J]. Butterworths, London.

[62] YALIN M S, 1983. River Bed Degradation, Downstream of a Dam [C]. Proceedings, XX Congress of IAHR, Moscow, USSR.

[63] CARSLAW H S, JAEGER J C, 1973. Conduction of Heat in Solids [J]. Clarendon Press, Oxford.